Electricians' On-Site Companion

'This book is must for all electricians to have in their toolkit. It gives good practical advice on how to tackle real-life situations and doesn't just focus on how to pass an examination.'

— **David Nunan**, *Steve Willis Training Ltd.*

This book contains everything electricians need to know about working on site, covering not only the health and safety aspects of site work, but also the techniques and testing knowledge required from the modern-day electrician. Regulations issues are included alongside step-by-step instructions for each task, after which testing information, checklists and example forms are given so that site workers can ensure they have done everything required of them.

Christopher Kitcher has been working in the electrical industry for over 50 years. As well as having been both a contractor and a builder, for many years Chris was a Lecturer in Electrical Installation at Central Sussex College and is still an NICEIC inspector for the Microgeneration Certification Scheme (MCS). For the last 17 years he has worked in the college environment while maintaining his electrical skills by periodically working on site.

Electricians' On-Site Companion

Christopher Kitcher

Routledge
Taylor & Francis Group

LONDON AND NEW YORK

First published 2018
by Routledge
2 Park Square, Milton Park, Abingdon, Oxon OX14 4RN

and by Routledge
711 Third Avenue, New York, NY 10017

Routledge is an imprint of the Taylor & Francis Group, an informa business

British Library Cataloguing-in-Publication Data
A catalogue record for this book is available from the British Library

Library of Congress Cataloging-in-Publication Data
Names: Kitcher, Chris, author.
Title: Electricians' on-site companion / Christopher Kitcher.
Description: New York, NY : Routledge, 2018. | Includes bibliographical
    references and index.
Identifiers: LCCN 2017009250 | ISBN 9781138683327 (pbk. : alk. paper) |
    ISBN 9781138309326 (hardcover) | ISBN 9781315544571 (ebook)
Subjects: LCSH: Electric wiring—Handbooks, manuals, etc. | Buildings—Electric
    equipment—Installation—Handbooks, manuals, etc. | Electric engineering—
    Insurance requirements—Handbooks, manuals, etc.
Classification: LCC TK3205 .K38 2018 | DDC 621.319/24—dc23
LC record available at https://lccn.loc.gov/2017009250

ISBN: 978-1-138-30932-6 (hbk)
ISBN: 978-1-138-68332-7 (pbk)
ISBN: 978-1-315-54457-1 (ebk)

Typeset in Sabon
by Apex CoVantage, LLC

Contents

Chapter 1

Design considerations

It is vital for a satisfactory outcome to any electrical installation that it is designed correctly. It doesn't matter whether the installation is a single circuit or a full installation, the same rules apply.

The statutory document that applies to all electrical installations is the Electricity at Work Regulations 1989 (EWAR 1989). These regulations are made under the Health and Safety at Work Act 1974 (HASWA 1974) and quite clearly state that although BS 761 wiring regulations are non-statutory, compliance with them will be likely to achieve compliance with the relevant parts of the EAWR.

The HASWA places a duty of care on employers, the self-employed and on employees. EAWR also places responsibilities on everyone at a place of work – this, of course, means employers, self-employed persons and employees.

An employer is any person or body that employs one or more persons under a contract of employment or apprenticeship, and it is the employer's duty to comply with the EAWR. It is the employee's duty to co-operate with their employer to enable the duty placed on the employer by the EWAR to be complied with.

Under normal circumstances, the person who has the electrical installation under their control is classed as the duty holder; this may apply to the whole installation or just the part of the electrical installation which is being worked on. Of course, if there is no work being carried out the duty holder would be the managing director or owner as it is his duty to ensure that the EAWR are complied with in any place of work.

The important thing to remember is that the EWAR applies to everyone wherever work is being carried out, and that the duties in some of the regulations are subject to the term 'reasonably practicable'. Where the term 'reasonably practicable' is not used, then the regulation is said to be absolute and must be adhered to.

When a person has to do something so far as reasonably practicable, they must assess the magnitude of the risk for the work activity against the cost in terms of physical difficulty, trouble, time and the expense that would be involved in eliminating or reducing the risk. The term 'reasonably practicable' would be acceptable where the risks to health and safety for a particular job are low, and the cost or technical difficulties to reduce the risk are high.

Where the risk is high and may result in death, then the level of duty would always be absolute and cost would not form part of the equation.

When designing any electrical installation, all regulations that have or may have an effect on it must be taken into account. None of these regulations is more important than the others, although of course some may be more relevant.

As an example, the Health and Safety at Work Act will always form part of any electrical design but Part P of the building regulations will not apply to buildings other than dwellings.

All of the relevant statutory regulations are in place to ensure that electrical installations are installed safely and remain safe; one of the major requirements is that materials and equipment are manufactured and installed to a high standard. During the design stage it is important that we ensure that all materials that are intended to be used in the installation are to a British or harmonised standard.

BS 7671: 2008 has been adopted as the British Standard for electrical installations; it is not a statutory document but compliance with this standard will ensure compliance with the other statutory regulations. Chapter 51 of BS 7671 requires that all equipment complies with a British or harmonised standard.

This regulation is not intended to prevent the use of non-BS equipment, but it makes it quite clear that any equipment that does not have a British or harmonised standard must be compared with the standards and that any differences must be verified by the designer or installer. Any differences must not result in a lesser degree of safety, and the use of the equipment must be recorded on the inspection and test documentation that is completed for the work.

An example would be where a wrought-iron light fitting has been made by the local blacksmith. This would not have a BS, however it could be installed providing the person installing it was happy that it was safe, and that the correct documentation was completed.

If it was just the installation of a light fitting then it would be recorded on a minor works certificate (Figure 1.1) in Part 1, Item 4, which is headed as 'Details of departures', provided, of course, that the fitting was being connected to an existing circuit. New inventions that have not yet been given a British Standard would also need to be entered as a departure.

BS 7671 consists of 7 parts and 16 appendices; each part is broken up into chapters.

Each regulation has a number, and in the system used the first digit is the part number, the second digit is the chapter and the third digit is the section; the remaining digits are the actual regulation number.

Megger.

MINOR ELECTRICAL INSTALLATION WORKS CERTIFICATE
(REQUIREMENTS FOR ELECTRICAL INSTALLATIONS - BS7671 [IET WIRING REGULATIONS])
To be used only for minor electrical work which does not include the provision of a new circuit.

PART 1: DESCRIPTION OF MINOR WORKS

1. Description of the minor works
2. Location/Address

3. Date minor works completed
4. Details of departures, if any, from BS7671: as amended

5. Details of permitted exceptions (Regulation 411.3.3). Where applicable, a suitable risk assessment(s) must be attached to this Certificate.

PART 2: INSTALLATION DETAILS

1. System earthing arrangement   TN-C-S   TN-S   TT
2. Method of fault protection
3. Protective device for the modified circuit   Type   Rating   A

Comments on existing installation, including adequacy of earthing and bonding arrangements (see Regulation 132.16):

PART 3: ESSENTIAL TESTS

Earth continuity satisfactory

Insulation resistance: Live-Live $M\Omega$

                   Live-Earth $M\Omega$

Earth fault loop impedance Ω

Polarity satisfactory

RCD operation (if applicable).   Rated residual operating current ($I_{\Delta n}$)   mA

Disconnection time at $I_{\Delta n}$ ms

Disconnection time at $5I_{\Delta n}$ ms

Satisfactory test button operation   (Insert ✓ to indicate operation is satisfactory)

PART 4: DECLARATION

I CERTIFY that the said works do not impair the safety of the existing installation, that the said works have been designed, constructed, inspected and tested in accordance with BS 7671: 2008 (IET Wiring Regulations), amended to and that the said works, to the best of my knowledge and belief, at the time of my inspection, complied with BS7671 except as detailed in part 1 above.

Name: Signature:

For and on behalf of: Megger

Address: Archcliffe Road

     Dover   Position:

     Kent   Date:

Figure 1.1

Example: regulation 421.1.1

This will tell us that the regulation is Part 4, Chapter 2, section 1. This covers the General Requirements of this chapter. Then we have the 1.1, which shows us that it is the first regulation in that section and is the first part of that regulation.

Any regulation ending with a number in the 200 range is specific to the UK.

Example: regulation 422.4.203

This tells us that the regulation is Part 4, Chapter 2, section 2. It is regulation 4, which covers combustible construction materials and is section 203 of the actual regulation. This is specific to UK installations only.

For all electrical installations the basic requirements are the same:

- It must be designed to provide protection of persons, livestock and property.
- It must ensure the proper functioning of the electrical installation.

Before any design can begin, the following basic information about the installation will be required; it will also need to be entered onto the installation documentation on completion of the work.

Characteristics of the supply that is available

i. Is it AC or DC?
ii. Purpose and number of conductors:

- o Is the installation single phase or three phase and neutral?
- o Is there a supplier's earth connection?
- o Is there a PEN conductor (TN-C-S)?

iii. Values and tolerances:

- o Nominal voltage available, voltage tolerances. For a supply from the DNO the voltage tolerance is +10% and –6%
- o Maximum current that is available
- o Prospective short circuit current (I_{sc})
- o External earth fault loop impedance (Ze)

iv. Has the DNO any particular requirements?

Nature of demand

Once it has been decided that the supply is suitable for the requirements of the installation the next step is to decide on the types of circuits required.

This will need to include anything within the proposed installation that will require an electrical supply.

Items that need to be considered are:

   i. The location of points that will require a supply
  ii. Loads that will be expected on circuits
 iii. Variation of demands due to weekend use and seasons
 iv. Special conditions such as harmonics and start up currents
  v. Any requirements for information technology, controls and signalling
 vi. Any anticipated future demand due to additions and alterations

External influences/environmental conditions

Will the installation be subjected to any unusual conditions that will need to be taken into account during the design stage? These could be any that are listed in Appendix 5 of BS 7671.

Environment

Consideration should be given to the following:

Ambient temperature

Any high or low ambient temperatures that may be within the installation in places such as boiler houses, food preparation and cold storage areas.

Along with higher than usual temperatures humidity could also be a cause for concern as it could promote corrosion or dampness.

Water and corrosion

The presence of water whether it is between negligible or total submersion needs to be considered. Any equipment used will need to be capable of preventing the ingress of water and should be selected using the correct IP codes. As an example, external sockets and switches should be to IP56 and, of course, should be corrosion resistant. Corrosion could also be caused by chemicals either in liquid, powder or atmospheric form; these could be present continuously or only now and then.

Foreign bodies

In most cases this would be the presence of dust. It may be negligible but on occasion could be very heavy. Consideration must be given to the type of use that the building is being put to. Joinery workshops and granary stores are good examples of installations that may require a greater degree of protection.

Impact and vibration

All parts of all installations must be selected and erected to prevent damage by impact, abrasion, penetration, tension or compression. The type of protection will depend on the installation. As an example, cables buried in walls or below floors may require RCD protection or mechanical protection; this does not mean that cables clipped directly to a surface do not require protection. Where flat profile PVC cables are clipped on the surface in a loft space supplying a light or power point, it would be unlikely that additional mechanical protection would be required. This is because the outer sheath of PVC is classed as mechanical protection. It is also possible for this type of cable to be buried directly in the wall and plastered over with no further protection other than an RCD. It is usual, though, to protect the cable with capping or conduit; this is really just to protect it from damage that may be caused by a plasterer's trowel. When steel capping is used it is not a requirement for it to be earthed, as once it is plastered over it really is just a piece of metal.

I have noticed that on occasion some installers use the cpc to earth the metal capping and then consider it as earthed mechanical protection. This, of course, would not comply with BS 7671 as all electrical connections need to be accessible for future inspection, and would not be at all practical as galvanised capping has no provision for the connection of conductors.

On occasions when cables are to be buried in walls and for some reason RCDs are not to be installed, then the use of robust mechanical protection that will prevent the penetration of nails/screws can be used. Under these conditions, the protection need not be earthed. Usually, it is easier to use a less robust material and ensure that it is earthed; this would usually form or be part of the sheath of a cable.

The types of cable that would be suitable for burying directly into a wall could be: steel wire armoured, FP 200 Gold, Mineral, Flexishield (see Figure 1.2).

All of these cables are easy to terminate and will provide the required protection if installed correctly.

There are many ways in which protection from impact and vibration can be provided; this will depend on the type of use and level of protection required. In most cases there will be more than one method of protection that will be suitable. It is important to give some consideration to the appropriate type of protection. As previously mentioned, the outer sheath of flat twin and earth cable can be classed as mechanical protection (Figure 1.3), and is suitable protection for cables clipped direct to various surfaces, particularly where there is no risk of impact during normal use. In practice, it is often easier to use a piece of plastic conduit or mini trunking as it can often be quicker and the end result may look better.

If the installation is commercial or industrial, then, of course, the use of steel conduit and trunking should be considered along with all of the other types containment systems and cable types. This would include the use of busbar trunking systems, dado trunking systems and the like.

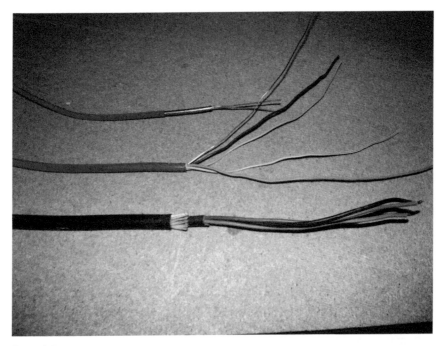

Figure 1.2

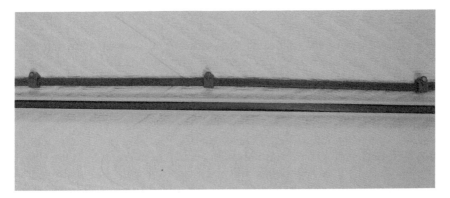

Figure 1.3

Vibration can sometimes be a problem and then there are various methods that can be used to prevent damage. One of the most common causes for concern with regards to vibration is the connection to motors and equipment with moving parts.

All methods of connection need to be thought about carefully because often it is not only vibration that needs to be considered. Other external

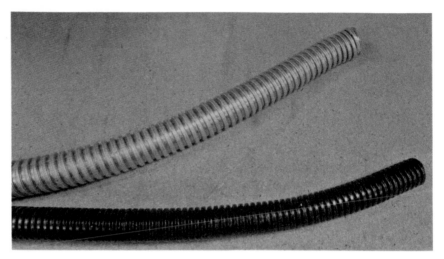

Figure 1.4

influences may need to be considered as well, which could be anything from corrosion, dust and impact, etc., as well as the adjustment of belts and chains, so it is important that the correct method is chosen.

In many cases, vibration and adjustment can be accommodated by the use of flexible conduit (Figure 1.4).

Flora and fauna

Consideration must always be given to any plant life or animal life that may exist in a particular region. All electrical installations must be protected from damage that may be caused by the growth of plants or the activity of animals that may be present.

As an example, the growth of ivy around a conduit installation may well cause damage a few years down the line; the ivy could wrap around the conduit and eventually force it off of the wall. In this instance, it would not be wise to use plastic conduit as it may bend and crack. Where steel conduit is to be used the fixings must be of good quality and well fixed; if in doubt, put the fixings closer together for added support. Whenever we install equipment we must assume that it is going to be in place for many years and thought must be given to anything that could possibly cause damage over a prolonged period.

'Fauna' is a term given to animal life, especially those of a particular region or area. Fauna is, of course, more difficult to provide protection for as creatures like mice, rats and squirrels can get into most buildings with very little effort.

Any electrician will tell you that these creatures love to chew on cable insulation; it would be impossible to protect every cable from a mouse that wants to chew a cable. However, where there are a lot of mice, such as a grain store on a farm, precaution against damage from mice is a must. Care must be taken to ensure that they cannot get into the cable containment system by making sure that all equipment meets the required IP code and that all lids are secure.

Other problems can be caused by the ingress of ants making nests in the base of lamp posts and conduit boxes; all of these things should be considered as indicated in Section AK and AL in Appendix 5.

An IP code is an international protection marking that is designed to classify the levels of protection against the intrusion of bodily parts, dust, solid foreign objects and liquids.

Table 1.1a sets out the IP codes that are commonly used in BS 7671.

Table 1.1a

Degree of protection against the ingress of dust and foreign objects − 1st number		
First number	Level of protection	Definition
0	No protection	
1	No penetration by an object of 50mm diameter or greater	50mm ball shall not pass through hole
2	No penetration by an object of 12.5mm or greater	12.5mm ball shall not pass through hole
3	No penetration by an object of 2.5mm diameter or greater	2.5mm ball shall not pass through hole
4	No penetration by an object of 1mm diameter or greater	1mm ball shall not pass through hole
5	Dust protected	Dust protected enough to prevent any dust affecting equipment operation
6	Dust proof	No ingress of dust possible
Degree of protection against the ingress of liquids − 2nd number		
Second number	Level of protection	Definition
0	Not protected	
1	Vertically falling water	Vertically falling drops of water will cause no harm
2	Vertically falling water when enclosure tilted up to 15°	Vertically falling drops of water will cause no harm
3	Spraying water	Vertically spraying water at an angle of 60° from vertical will cause no harm

(Continued)

Table 1.1a (Continued)

Degree of protection against the ingress of liquids – 2nd number

Second number	Level of protection	Definition
5	Jets of water	Jets of water from any direction will cause no harm
6	Powerful Jets of water	Powerful jets of water from any direction will cause no harm
7	Temporary immersion	Temporary immersion to a depth of 1m will not result in harmful effects due to water penetration
8	Continuous immersion	Continuous immersion will not result in harmful effects due to water penetration

First additional letter

A		Protected from access by the back of a hand
B		Protected from access by a British Standard finger 12mm in diameter and 80mm long
C		Protected against access with a tool
D		Protected against access by a wire probe 100mm long

The code used in BS 7671 consists of two digits and a letter; the first digit indicates the level of protection against the ingress of dust and solid foreign objects.

The second digit indicates the level of protection against the ingress of liquids.

The letter indicates the level of protection from specific items.

Where an X is used, it means that there is no protection. As an example, the front surface of an electrical enclosure installed indoors has to comply with the minimum of IP2X. This indicates that penetration by a sphere 12.5mm in diameter will be prevented. Clearly, the hole could be 12mm in diameter but it will not prevent the ingress of liquids. However, as the enclosure is indoors the presence of liquids is unlikely. For that reason, water protection is not required and an X is used.

Electromagnetic influences

Many things can be the cause of electromagnetic disturbances. These can include the switching of:

- Electric motors
- Inductive loads
- Transformers
- Discharge lighting
- IT equipment

Where there is a possibility of damage or interference being caused by voltage surges or electromagnetic disturbances, precautions should be taken. Consideration should be given to the use of:

- Surge protection devices
- Screened data cables with bypass conductors fitted
- An equipotential bonding network where the following are connected to the network:

    (i) Metal containment systems, conductive sheaths and armouring of data transmission cables or information technology equipment
    (ii) Functional earthing conductors
    (iii) Protective conductors

Where equipment such as information technology equipment contain components that operate at high frequency and use a functional earth, it is important to ensure that the bonding conductors have a very low impedance. This can be achieved by the use of individual bonds, which need to be kept as short as possible, where possible consideration should be given to the use of a bonding grid such as described in Annex A444 of BS 7671.

Utilisation

Consideration must always be given to what the building is being used for, as obviously this has an effect on the installation for many reasons. One of the first things to consider is the **capability** of the persons using the installation.

Where the installation is to be used by **ordinary persons** then all that is required is that the system will operate safely under any external influences that may be present.

Where it is known that **children** are going to be present, higher levels of IP rating should be required; any equipment such as heaters with a high surface temperature exceeding 80°C should be made inaccessible to touch.

If areas of the installation are to be used by **disabled** persons, the characteristics of the installation should be suitable for the nature of the disability. As an example, switches and sockets may need to be made inaccessible to prevent misuse.

Additional consideration would need to be given in areas such as prisons or unattended installations such as illuminated bus stops, telephone booths, town maps. These areas may be open to vandalism and perhaps the system would be better if it was completely contained within the building structure, or where it has to be surface mounted, tamper-proof screws and additional fixings should be considered.

Perhaps the installation will allow the user to be in **contact with earth**, which, of course, will increase the risk of electric shock. This is the reason that some installations are subject to additional requirements by BS 7671; examples of this would be construction sites and agricultural installations.

Will the contact with earth be continuous or will it be low? Where it is continuous, then converting TN-C-S and TN-S systems to TT may be considered, along with using non-conductive containment systems. If the risk of contact with earth is nonexistent, such as a second-floor flat, then there would be no reason to consider any special precautions.

Regardless of the type of installation, consideration must always be given to **evacuation** of the area in the event of an emergency.

In any area that may form part of an escape route, methods must be in place to prevent the collapse of any part of the electrical system that may block the route. Plastic trunking or conduit must be fixed with metal restraints to prevent it falling down in the event of a fire (reg. 521.11).

In areas where there are difficult conditions for evacuation or high density occupation, consideration must also be given to using flame retardant and low smoke emission materials (reg. 422.1)

The storage and use of **materials** has to be taken into consideration. In areas where combustible materials are used or stored there will always be an increased risk of fire. As an example, a carpenter's workshop will have wood and wood shavings scattered around. This is, of course, a fire risk.

Where particular types of adhesives are used or perhaps fuel/gases are stored, there will be a risk of explosion. There will also be a risk of explosion where there is a build-up of dust. Under these conditions it is important to ensure that the materials used for the installation are suitable. The dangerous substances and explosive atmospheres regulations 2002 should be used to decide the level of risk and as a minimum any testing carried out should be intrinsically safe.

Due consideration must also be given to any materials/substances that may cause contamination; anything from powders to acids may present a problem and the parts of the installation that may be affected must be designed around them.

Construction of buildings

Fire propagation

The construction of a building will always have to be taken into consideration when designing an electrical installation. The system must be designed taking into account the type of construction materials used.

The first consideration would always be fire. Clearly, timber-framed buildings or perhaps even large tents or marquees would be considered combustible and the materials used for the electrical installation must be selected to provide the best protection.

One source of information when working in marquees is a guide that is produced by the Performance Textiles Association. Their trading name is

MUTA and they produce a best practice guide, which can be found on their website (www.mutamarq.org.uk).

Where installations are near combustible construction materials, regulation 422.4 should be applied; this regulation states that measures must be taken to ensure that electrical equipment cannot cause ignition of walls, floors or ceilings.

Where pre-fabricated buildings are being erected, any boxes and enclosures that are installed in premade walls must have protection to at least IP3X where the wall is likely to be drilled during erection. This is to prevent any dust/shavings finding their way into the enclosure and creating a possible fire risk.

Precautions must also be taken to ensure that all enclosures installed on or in combustible materials will not affect the surrounding materials due to any temperature rise caused by the operating of the installation. It is always worth remembering that parts of the electrical installation can safely operate at temperatures of up to 70°C.

Of course, it is not only the temperature of the installation that needs to be taken into account, as other equipment such as heaters and lamps could also cause a problem if they are installed without giving due consideration to where they are sited.

Lamps, such as halogen ones, should have a protective screen to prevent and hot particles falling on, or being projected onto any combustible surfaces. The light source of a lamp should also be kept at a suitable distance from any surface that may be affected by the heat emitted from it.

Where manufacturers' guidance is not available, regulation 422.3.1 gives some guidance but common sense must also be applied, and consideration must be given to:

• The maximum power likely to be dissipated by the lamp.
• The type of material the lamp is fitted to or where the heat from the lamp is directed. Is it fire resistant?

As an example, the regulation uses a spotlight or projector. But, of course, some types of luminaires may get hotter than the examples given and in these cases the distances should be increased. The example distances provided are shown in Table 1.1b:

Table 1.1b

Maximum rating in watts	Minimum distance from surface
Up to 100W	0.5m
Over 100 and up to 300W	0.8m
Over 300 and up to 500W	1.0m

Table 55.3 in BS 7671 provides a list of symbols with explanations that should be found on any British standard light fitting.

The first three symbols shown are the ones used prior to the latest edition of BS EN 60598-1: 2008. The last three symbols are from the latest edition and you will see from the descriptions that only luminaires that are not permitted to be fitted are marked.

Fittings with these symbols indicate:

The luminaire is suitable for mounting on normally flammable surfaces. However, you need to be aware that the latest edition of BS EN 60598-1 does not require this type of fitting to be marked at all. Only luminaires that must **not** be fitted to combustible surfaces need to be marked.

The luminaire can be mounted on normally flammable surfaces and also be covered with thermally insulating material if required.

The luminaire must not be mounted any combustible surface.

A recessed luminaire that is not suitable for mounting onto normally flammable surfaces.

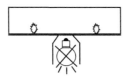

A surface luminaire that is not suitable for mounting onto a normally flammable surface.

A luminaire that is not suitable for covering with thermally insulating material.

Structural movement

Many buildings are subject to structural movement and, of course, when installing an electrical system this has to be taken into account. Where long runs of conduit or trunking are installed along walls, the expansion of the wall will be different to the expansion of the steel or plastic containment.

Where equipment is fixed and is reasonably solid, consideration must always be given to expansion, particularly over long runs that may be subjected to temperature changes. Where plastic conduit is to be used a special coupler should be used that has an elongated socket that will allow the conduit to expand and contract without compromising the containment system.

The expansion coupler is glued at the short end and the long end is left as a push fit. Always be aware of the temperature during installation when using them. If it is cold, allow some room for expansion; if it's hot, then allow some room for contraction (see Figure 1.5).

If it is cold, there would be little point in pushing the conduit into the coupler right down to the stop, as the whole idea is to let the conduit expand as it heats up. Usually, if the conduit is around halfway into the coupler it will be about right whatever the temperature.

For an installation where steel is used the solution is not quite so simple. On a long run of conduit that is exposed to major changes in temperature, it is important to allow for expansion. Usually, there is enough play in the saddles to allow the tube to move. However, attention should be paid to the fixings as the movement may put a bit more stress on them and instead of expanding the tube may try and buckle.

To allow for expansion of the tube it is often a good idea to put a U shape in it, which will serve a double purpose. Not only will it allow the tube to

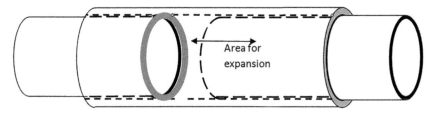

Area for expansion

Figure 1.5

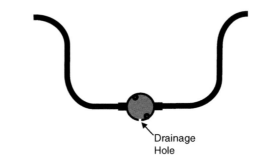

Drainage
Hole

Figure 1.6

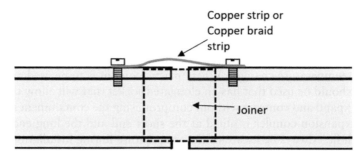

Copper strip or
Copper braid
strip

Joiner

Figure 1.7

expand, it will provide a low point to allow any water that has found its
way into the conduit system to drain out. Of course, you will need to drill a
hole in it at its lowest point. It's not a bad idea to put a through box in the
bottom of the U; this will make the pulling of cables easier and it can have
a hole drilled in the bottom of it instead of drilling the conduit (Figure 1.6).

Trunking can be dealt with by extending the slots in the trunking where it
joins to the trunking connector and drilling slightly larger fixing holes and
using screws and washers to allow for expansion. In situations where the
joints need to move then it is important that the integrity of the earth continuity of the trunking is maintained; this can be achieved by fixing a flexible
lead across the joint (Figure 1.7).

Chapter 2

Types of supply system and earthing arrangements

We have three types of system that are in common use and they are referred to as TT, TN-S and TN-C-S.

Where the letter T is used when identifying supply systems, it represents earth. T is used because it is the first letter of 'terra', which is the Latin word used for earth, dirt or land.

TT supply system

A TT system is normally a supply in which the cables are run into the building above ground. The abbreviation TT indicates that the supply system and the installation have two points of earth.

In many rural areas and some urban areas you will see electricity cables suspended above ground by poles that look exactly the same as telegraph poles, the difference being that they usually have four wires suspended from them, one above the other.

In most domestic properties two cables are run into the house; these are the line and neutral. In this type of installation the earth for the property is not supplied by the district network operator (DNO). As we know an earthing system is vital to the safety of most installations. Where we have a TT system, the earth has to be provided by the installer of the electrical installation. This is carried out by using an earth electrode which is dedicated to the installation.

An earth electrode is driven into the ground in a convenient position, and an earthing conductor is connected to it and run into the consumer unit; this will provide the earth for the property (see Figure 2.1).

As you can see from Figure 2.1, when there is an electrical fault to earth, the fault current will flow through the electrical supply line conductor to the consumer's electrode and then through the earth (soil) back to the supply transformer electrode. The fault current will then flow through the transformer winding to the origin of the supply.

A TT system can usually be identified at the service head (Figure 2.2). This is because there will be no sign of an earthing conductor being connected to

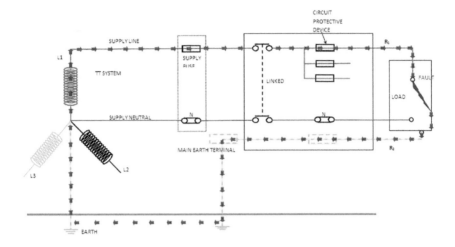

Figure 2.1

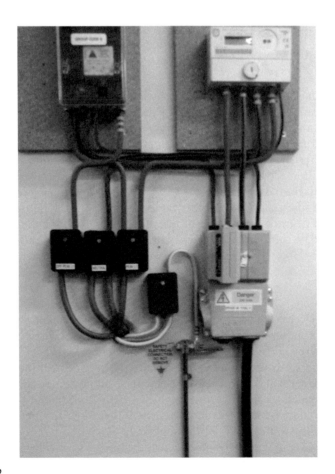

Figure 2.2

the service head, or the supply cable. Most, if not all, TT systems require an RCD to be installed to ensure that disconnection times are met in the event of an earth fault.

The resistance of the suppliers earth electrode is stated as being 21Ω but, of course, this value cannot be stated as Ze because the resistance of the consumer's earth electrode and the resistance of the earth between both the distributor's electrode and the consumer's electrode has to be taken into account. Further explanation of this is provided in the earth electrode section.

TN-S supply system

A TN-S supply system uses only one point of earth and this is supplied to us by the DNO. As can be seen from Figure 2.3, the point at which the system is connected to earth is at the supply transformer. In the event of a fault, the current can flow through the sheath of the supply cable or perhaps a separate earthing conductor. The abbreviation TN-S is earth and neutral separate; this is because the earth and the neutral are completely separate throughout the installation.

This type of supply is usually taken into the house underground. The cable is lead covered and protected by a steel sheath that is covered in hessian. A connection is made to the lead sheath by the DNO and this provides the earth for the system (Figure 2.4). In most installations where the supply is provided through a TN-S system the resistance of the fault path is quite low with a typical maximum Ze of 0.8Ω.

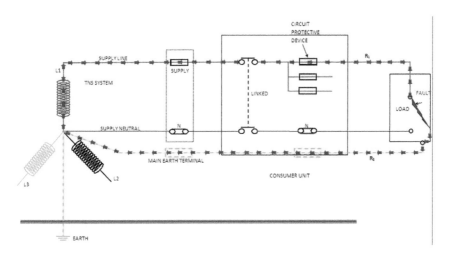

Figure 2.3

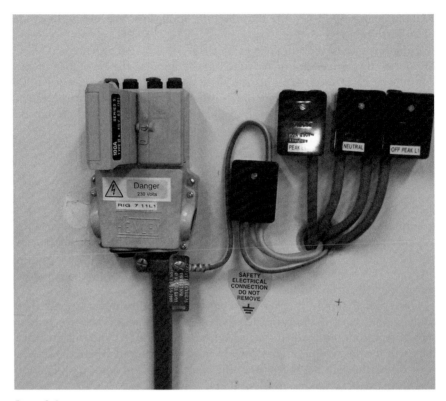

Figure 2.4

TN-C-S supply system

A TN-C-S system also uses a single point of earth. The difference between this system and the TN-S system is that the supply neutral in a TN-C-S system serves a dual purpose; not only is it used as the neutral conductor, it is also used as the supplier's earth (Figure 2.5). In this system the supply neutral is known as the PEN conductor (protective earth and neutral).

In modern houses this type of supply would be run into the house underground. However, there is no reason why it could not be overhead and, in fact, many TT systems have been converted to TN-C-S systems. As you can see by the drawing, the earthing of this system is provided by connecting an earthing conductor to the supply neutral at the service head (Figure 2.6). This connection must be made by the DNO and must not be made by anyone else.

TN-C-S systems are often referred to as PME systems (protective multiple earthing).

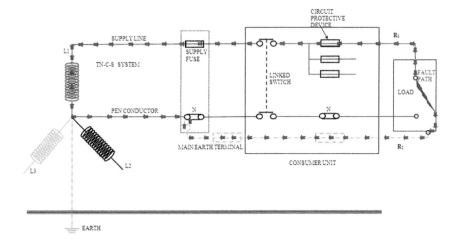

Figure 2.5

Figure 2.6

Earthing

To prevent damage to property, or injury to persons or livestock when there is a fault to earth, it is a requirement that the system is installed so that the supply to the fault is interrupted very quickly. To ensure that this is the case, the supply system and each circuit must be co-ordinated correctly to ensure that the protective device will operate when required.

This relies on the resistance of the earth fault path being low enough to allow the amount of current to flow that is required to operate the protective device in the required time.

In the first instance it is necessary to select the correct size of the supply earthing conductor. Where the system is a TN-S system, the earthing conductor must be a minimum of half the size of the meter tails.

Example:

- 16mm^2 tails = 10mm^2 copper earthing conductor
- 25mm^2 tails = 16mm^2 copper earthing conductor

For a TN-C-S supply a 16mm^2 copper earthing conductor should be used for all installations with meter tails up to and including 35mm^2.

TT systems are dealt with differently because they use an earth electrode and an RCD; the earth fault current will be a lot lower than that which would occur in the other systems.

Providing the earthing conductor is not buried and is protected against corrosion and mechanical damage it can be 2.5mm^2. This would require the conductor to be enclosed in mini trunking or conduit. In cases where the conductor is not protected it can be 4mm^2 copper.

In some instances where the earth electrode is a distance from the building, it is necessary for the conductor to be buried in the ground. In these situations the conductor can be 2.5mm^2 providing it is protected against corrosion and mechanical damage.

Where the conductor is protected against corrosion only, then it must be a minimum of 16mm^2. If it is not protected against corrosion, then the minimum size is 25mm^2.

Chapter 3

Earth electrodes

Earth electrodes

The most common use of an earth electrode is for providing an earth for a TT system.

A copper-plated steel rod is the most common type of earth electrode for use in a domestic installation; a steel pipe can also be used. Earth plates and underground structural metalwork can also be used where suitable.

Before installing an earth electrode, consideration must be given to where it is going to be positioned. Ideally, it should be placed away from areas where it is likely to get damaged or to come into contact with chemicals that may cause it to corrode. It must also be in an area where the soil is unlikely to freeze.

An ideal area is one that is in the shade and where the soil has less chance of drying out. A standard rod electrode used in a domestic installation would be driven into the ground to a depth of around 1m. Ideally, the electrode will be installed in an inspection pit (Figure 3.1). In most cases a lightweight plastic enclosure can be used.

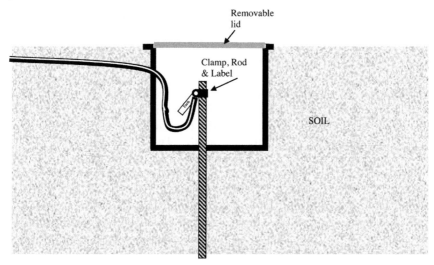

Figure 3.1

Where an earth electrode is used for protection on a TT system, it is important that the electrode provides a low enough resistance to allow enough earth fault current to flow. To operate a 30mA RCD a fault current of 30mA is required. To allow 30mA to flow through the earth fault path the resistance of the earth fault loop must be no greater than $\frac{50}{0.03} = 1667\Omega$.

Within a TT installation the touch voltage under fault conditions must never be permitted to rise above 50V. To calculate the maximum resistance of the earth fault loop path we can use Ohm's law and divide the voltage by the current flow in amperes.

As the trip rating of the RCD used is given as 30mA we need to divide 30 by 1000 to convert it to amperes. We can see that the maximum earth permitted earth fault path is 1667Ω.

The problem with using this value is that it will be unstable. Due to the nature of soil, the resistance will change almost daily.

In dry weather the resistance will be higher than it will be in wet weather. If we were to test the resistance of our electrode on a wet day it may have a low enough resistance to make it suitable for use. If the soil were to dry out, as it would in the summer, the resistance will rise, and it may rise above the permitted maximum value.

To ensure that the earthing system remains effective, the maximum resistance of the earth fault loop is limited to 200Ω. Any resistance above this will be considered unstable.

Circuit protective conductors (cpcs)

The resistance of the supply side of the installation is something that we have no control over. The symbol for the supply resistance, which is usually referred to as the external impedance, is Ze. This is because the resistance of an AC circuit is referred to as impedance, and the letter used for impedance is Z.

All final circuits that have a current rating of up to and including 32A must automatically disconnect from the supply in the event of a fault to earth within 0.4 of a second. This requires that the impedance (resistance) of the circuit must be low enough to allow the required amount of current to flow. The value of resistance for each circuit is known as Zs and this is made up of the resistance of Ze plus the resistance of the circuit line (R_1) and circuit protective conductor $(R_2$ earth). The formula used for Zs is: Zs = R_1 + R_2 and Figure 3.2 shows how this value is made up.

As you can see, the value of Ze is made up of the resistance of the supply line and the resistance of the path that the fault current would take back to the supply transformer; this is known as the earth fault return path. As seen in the previous drawings, in a TT system the earth fault path is through the mass of earth. For calculation purposes, this is given a value of 21Ω,

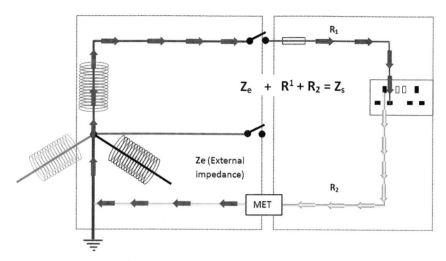

$$Z_e \;+\; R^1 + R_2 = Z_s$$

Ze (External impedance)

R_1

R_2

MET

Figure 3.2

although, of course, in reality it could be anything as the true resistance will depend on the type and size of the earth electrode used, along with the type of soil and the soil condition, which will depend on the weather.

A TN-S system that uses the metallic sheath of the supply cable will have a maximum Ze value of 0.8Ω and a TN-C-S system will have a maximum Ze value of 0.35Ω.

To ensure that enough current can flow to operate the protective devices, the circuit cables that we install must be very carefully calculated to prevent the Zs value from exceeding the permitted value.

Tables are provided for us that show the maximum permitted value of Zs for the type and rating of the protective device that is being used; it is up to us to use the correct size of cable to ensure that these values are not exceeded. Circuits that are installed with a Zs value that is too high could result in electric shock, fire or cable damage.

A BS 3036 semi enclosed rewirable fuse (Figure 3.3)

The measured Zs value for BS 3036 fuses must not exceed those shown in Table 3.1.

BS 88 cartridge fuse and BS 1361 cartridge fuse

The measured Zs values for BS 88 cartridge fuses must not exceed those in Table 3.1.

Figure 3.3

Figure 3.4

Table 3.1

MCBs and RCDs

Rating of MCB in amps	3	6	10	16	20	32	40	50	63
Type B — Max measured Zs for both 0.4 and 5 seconds	11.65	5.87	3.5	2.2	1.75	1.1	0.88	0.7	0.56
Type C — Max measured Zs for both 0.4 and 5 seconds	5.82	2.91	1.75	1.09	0.87	0.55	0.44	0.35	0.56
Type D — Max measured Zs for both 0.4 and 5 seconds	2.91	1.46	0.88	0.55	0.44	0.28	0.22	0.18	0.28

Rated residual operating current	Nominal voltage				
	50–120V	121–230V	231–400V	>400V	Max Z^s
30	1667*	1667*	1553*	1667*	
100	500*	500*	460*	500*	
300	167	167	153	167	
500	100	100	92	100	

*Earth electrode resistances below 200Ω may be unstable

Rating of fuse in amps	5	6	10	15	16	20	25	30	32
BS 1361 — Max measured Zs for 0.4 seconds	8Ω			2.5Ω		1.29Ω		0.7Ω	
BS 88 — Max measured Zs for 0.4 seconds	7.9Ω	6.2Ω	3.7Ω		1.84Ω	1.34Ω	1.09Ω		0.6Ω
BS 3036 — Max measured Zs for 0.4 seconds	7.3Ω			1.9Ω		1.3Ω		0.83Ω	

Use 0.4 seconds for circuits up to and including 32A and 5 seconds for above 32A

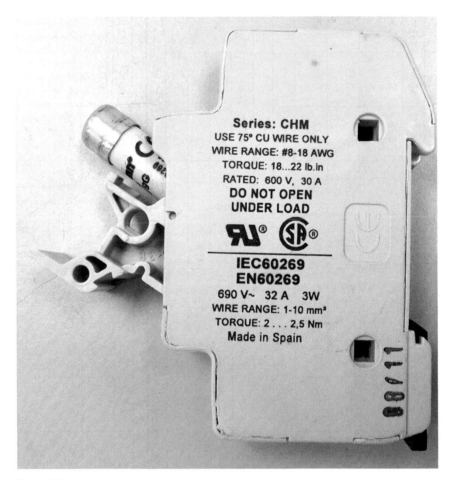

Figure 3.5

The measured Zs value of a circuit protected by a BS 1361 cartridge fuse must not exceed those in Table 3.1.

Figure 3.5 is of a fuse holder for a domestic cartridge fuse.

BS EN 60898 circuit breaker (Figure 3.6)

The measured Zs value for circuits protected by BS EN 60898 circuit breakers must not exceed those in Table 3.1.

Let's look at how an electric shock could occur if the earth was not correctly connected or the circuit resistance to earth (Zs) was too high. In Figure 3.7 we have an electric cooker that has become live due to a fault on one of its oven elements. For some reason, the earthing conductor has not been connected.

Figure 3.6

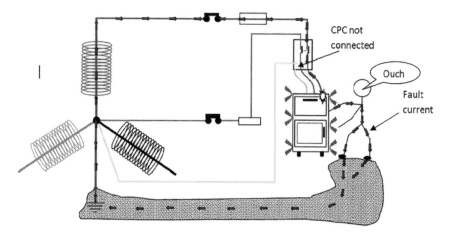

Figure 3.7

In this situation the case of the cooker will become live, although the elements that are not damaged will continue to work correctly. If you were to touch the cooker at the same time as you touched the water tap, there is a very good chance that you would receive an electric shock. This is because the tap will be at 0V and the cooker will be at a voltage considerably higher, probably 230V. As there is now a difference in pressure the current will be able to flow from the cooker through your body to the tap.

As you can see, earthing is very important as it is part of the safety system of an installation.

Selecting the correct size of circuit protective conductor

Before we can select the correct size of the cable that we will use, we need to measure the external earth loop impedance Ze. We have already seen previously in this chapter that the maximum values for each system are provided, and if we need to we can use these values for calculation purposes if it is difficult to measure the value before the circuit is installed.

- TT $=21\Omega$
- TN-S $= 0.8\Omega$
- TN-C-S $=0.35\Omega$

Once we know the Ze value of the system we need to calculate the resistance of the fault path that the current will take through the cable. This the $R_1 + R_2$ value for the total length of the cable used. Before we can calculate $R_1 + R_2$ for the total length we need to know the resistance per metre of the cable. Table 3.1b provides us with these values.

Table 3.1b Resistance per metre of copper conductors at an ambient temperature of 20°C in mΩ/mm

Conductor cross-sectional area mm²	Resistance per metre (mΩ/m)
1	18.10
1.5	12.10
2.5	7.41
4.0	4.61
6.0	3.08
10.0	1.83
16.0	1.15

As an example let's say we need to install a cable for a new radial socket outlet and the information which we have is:

Supply system is a TN-C-S with a Ze value of 0.35Ω.

Circuit protection is a BS EN 60898 circuit breaker 16A Type B.

The cable that we want to use is 2.5mm² with a 1.5mm² circuit protective conductor.

The cable is 17 metres long.

Using this information, it is quite a simple process to check that the circuit will meet the requirements of BS 7671 with regards to earth fault protection.

Step 1

We must now calculate the resistance for the length of cable. Using the table we can see that the resistance per metre given for a 2.5mm² copper conductor is 7.41mΩ per metre and the resistance of 1.5mm² is 12.1mΩ per metre. If we now add them together we can see that the resistance of our cable is 19.1mΩ per metre.

We are going to use 17 metres of this cable.

$$17 \times 19.51 = 331.7 \text{m}\Omega$$

This value is in milli Ω and we need to convert it to Ω by dividing it by 1000.

$$\frac{331.7}{1000} = 0.33$$

(We only need to use the first two decimal places.) This is the $R_1 + R_2$ value of our cable at 20°C.

When our cables are under load the temperature of the copper conductors increases. As the temperature rises, the resistance of the copper conductor also increases. It increases at a rate of 2% for each 5°C rise in temperature.

When we are using thermoplastic flat twin cable we expect it to reach 70°C if it is fully loaded. That is why it is called 70°C thermoplastic cable. If we are carrying out cable calculation properly we need to take this rise temperature and resistance into account.

For a new circuit we need to multiply $R_1 + R_2$ by 1.2. This will increase the value of resistance by 20%. In other words, it will show us what the resistance of the conductors will be at 70°C.

So:

$$0.33\Omega \times 1.2 = 0.396\Omega$$

It is better to round the value up rather than down or you can use the exact figure.

Step 2

Now we can add the $R_1 + R_2$ to the Ze, our supply, to find the total Zs of our circuit, as $Zs = Ze + R_1 + R_2$.

$$0.396 + 0.35 = 0.746\Omega$$

The total Zs of our circuit is 0.746Ω.

Step 3

We must now refer to Table 3.1 to see what the maximum permissible value is for our circuit, which is being protected by a 16A Type B circuit breaker.

As we can see, it is 2.73Ω. This is higher than the calculated value for our circuit, which means that it will be suitable.

Zs in a TT system

You may well be thinking that the previous calculation carried out on a TT system would result in a very high value of Zs, which, of course, would be correct. The same cable used for a circuit connected to a TT system would give a Zs value of:

$$Zs = Ze + R_1 + R_2 \quad 21\Omega + 0.396\Omega = 21.396\Omega$$

This value of Zs would make the use of the circuit unacceptable. This is because, due to the nature of a TT system, it has a high Ze to begin with.

With TT systems it is normal to have a high Zs value and for this reason we must use a device called a residual current device (RCD) for protection from earth faults. This device measures the amount of current that is flowing in the line conductor and the neutral conductor.

In a good circuit the current should be the same in both conductors. When there is a fault to earth this will result in an imbalance as some of the current will flow in the cpc. When this happens the RCD will operate and cut off the supply to the circuit.

The use of an RCD for protection is not restricted to TT supplies. In some instances it is very difficult to install a circuit that meets a low enough Zs value for the circuit protection used on any type of supply. In these instances, it is permitted to use an RCD to ensure a fast disconnection of the supply.

It is important that every effort is made to ensure a low enough $R_1 + R_2$ wherever possible.

Chapter 4

Protective bonding

As well as having earthing as protection we must also have protective bonding. The correct installation of protective bonding considerably reduces the risk of electric shock within an electrical system. As we have seen earlier, there must be an imbalance of pressure for electricity to flow. Protective bonding prevents an imbalance of pressure by joining all metal parts that could become live during a fault.

In Figure 4.1 we have an item of Class 1 equipment that has a fault to earth and is within an equipotential zone.

If the protective device did not operate because the fault current was not high enough, or perhaps the device was faulty or incorrectly installed, then everything that was connected to the main earth terminal would become live.

The correct installation of protective bonding will ensure that there will be no difference in potential between any exposed conductive parts and extraneous conductive parts within the installation as it creates an equipotential zone.

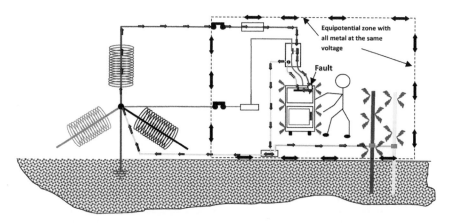

Figure 4.1

Everything within this zone will be live, but it will all be at the same potential and in this situation the current will not be able to flow from one live part to another. This will, of course, considerably reduce the risk of electric shock.

There are two types of protective bonding.

Main protective bonding is where a conductor is used to connect the main earthing terminal to extraneous conductive parts within the installation. Generally this would include:

- Metal water service pipes
- Metal gas installation pipes
- Oil service pipes installation
- Exposed structural steel of the building
- Metal central heating system
- Air conditioning system

Where any kind of bonding is carried out it is extremely important that it is installed correctly. We must always use the correct type of earth clamp to BS 951 (Figure 4.2), which must be fitted in the correct position within the installation. BS 951 earth clamps often have coloured markings; a red marking means that they must only be used in dry environments where condensation will not take place. These earth clamps should not be used to bond cold water pipes as they often get very wet with condensation.

The clamps that are coloured blue or green may be used in damp atmospheres.

The connection to the metal services entering the building should be on the consumer's side of the installation, within 600mm of the service meter. Where the service meter is external the connection can be made inside the building as near as possible after the point of entry before any branch in the pipe work. It is also permissible to make the bonding connection in the

Figure 4.2

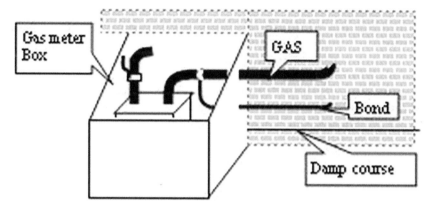

Figure 4.3

service meter box outside of the building; in these cases the cable must enter the box through its own dedicated hole (Figure 4.3).

Where the services entering the building are plastic, bonding must still be made to the metal installation pipes.

As a general rule, there is no requirement to bond plastic installation pipes, although in some installations where large pipes are used and the water has additives such as central heating protection protective bonding may be installed. This would require the use of metal inserts to attach the earth clamps to.

For TT, TN-S and TN-C-S systems there are calculations that can be made to select the size of main protective bonding conductors. The simplest method is to use 10mm² copper conductors as a minimum. Main protective bonding must have a resistance of 0.05Ω or less. In practical terms, this limits the length of a 10mm² copper conductor to 25m; where a length of more than 25m is required, then the size must be increased to 16mm.

Protective supplementary bonding is used to connect all of the extraneous conductive parts to all of the exposed conductive parts within a building. The use of protective supplementary bonding is often misunderstood; its function is to ensure that all of the metal parts within a certain area of an installation will rise to the same potential.

When there is an electrical earth fault within an installation, all of the metalwork that is connected in some way to the main earthing bar within the consumer's unit will become live. This, of course, includes any Class 1 equipment that is plugged in or connected to the installation in some way. In most cases the fault will operate the protective device instantly and there will not be a problem. However, if for some reason the protective device does not operate, not only will all of the metalwork connected to the earth terminal become live, any metalwork that is touching the earthed metalwork will also become live. This could include steel baths and bath taps, radiators, paperwork and, in fact, anything metal. Where they do become live it is likely that they will become live at a different potential, which could present a high level of shock risk, particularly in a room containing a bath or a shower.

To reduce the risk of electric shock protective supplementary bonding is installed. When installed correctly the supplementary bonding will ensure that the potential on all bonded metalwork will be the same, which of course will prevent any current flowing through a person who happens to touch two parts at the same time (Figure 4.4).

The difference between main protective bonding and supplementary protective bonding is as follows.

Main protective bonding is used to join all extraneous conductive parts within a building, such as the water and gas installation to the main earthing terminal. Protective supplementary bonding is used to connect all exposed and extraneous conductive parts within an area such as the bathroom. It is not necessary to connect this bonding conductor to the main earthing terminal, just the exposed and extraneous conductive parts together.

Where bonding is installed it is a requirement that all connections are made using the correct type of earth clamps to BS 951, and of course that the correct size of bonding conductor is used. Where the conductor is protected against mechanical damage by installing it in mini trunking or conduit the minimum size is 2.5mm^2. For a conductor that is clipped directly or run under flooring, then the minimum size is 4mm^2. In most instances, it is easier and cheaper to use 4mm^2.

The conductor used for all types of bonding must be identified by using green and yellow coloured cables just as you would for earthing conductors. Where the bonding is to be carried out in bathrooms that have copper plumbing the pipe work can be used as a conductor where required. This,

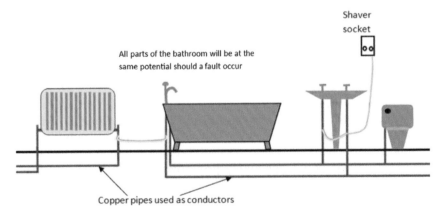

Figure 4.4

Figure 4.5

of course, is very helpful as the bonding conductors can be installed between the pipes supplying taps on the bath or/and basin only.

Bonding conductors must also be run from pipe work into any electrical points within the area being bonded. Remembering that the pipes can be used as bonding conductors, there is no problem with connecting from an electrical point to a pipe in the roof, as long as the pipe is connected to those being bonded.

Where you can be sure that the plumbing is all copper the bonding can be carried out in the airing cupboard simply by connecting a bonding conductor to each pipe (Figure 4.5).

In buildings that have plastic plumbing, or have RCD protection for all circuits in the bathroom, it is unlikely that bonding will be required.

Chapter 5

Diversity

In most, if not all, installations the full load will never be used and for that reason diversity can be applied not only to circuits to help us select the correct cable size. It can also be applied to whole installations to help us make a realistic assessment of the total amount of energy required from the supply.

I think that one of the best examples that could be used to explain the use of diversity is when a cooker is to be installed. It would be very unusual for a cooker to have all of its individual components drawing current at the same time.

Calculating the diversity of a cooker

When wiring a cooker point the same process should be followed as for all other circuits. Planning is very important and will save problems and time.

When we are installing a point for a cooker we need to ensure that the cable is a suitable size to carry the current required as cookers vary greatly in loading.

First, we must calculate the total load of the cooker. This is often given in the user instructions but if we can't find it we can always calculate it.

As an example let's take a cooker with the following:

- Two hob plates with a maximum output of   1.5kW each = 3kW
- One hob plate with a maximum output of   2kW = 2kW
- One hob plate with a maximum output of   1kW = 1kW
- One oven with a maximum output of   4.5kW = 4.5kW
- One small oven with a maximum output of   2kW = 2kW
- One grill with a maximum output of   2.5kW = 2.5kW

This will give us a total load of 15kW, which can now be converted to amperes by dividing by the voltage.

$$\frac{15000}{230} = 65.2\text{A}$$

This type of load is not unusual for a cooker; however, it is unlikely that in a domestic installation the whole of the load will be on at any one time, even on Christmas Day. This is because all of the elements are controlled by thermostats and all of the ingredients that require cooking require different cooking times.

Taking all of this into consideration, we can use something called diversity. This is something that will be explained in detail and a cooker is a good place to start.

Because not all of the loads of the cooker will be on together, we can carry out a simple diversity calculation to take this into account.

From the total load of 65.2A we can use 10A.

This leaves $65.2 - 10 = 55.2A$

Of the current that is left we can use 30%.

$$55.2 \times 30\% = 16.56A$$

Now, we must add to the 16.65A the 10A that we subtracted, which equals 26.65A, and this is the total load that we need to allow for the cooker circuit.

This is why most cooker circuits are rated at 32A and are wired in 6.00 mm² cable.

Some cooker outlets have a 13A socket outlet incorporated in them. In these cases, another 5A must be added to the total, which in this case is $26.65 + 5 = 31.65A$. This, of course, still allows us to use a 32A protective device.

This calculation can be used for all domestic cooker installations.

It is not unusual for the oven to come prefitted with a 13A plug top and the hob to be a separate piece of equipment. In these instances, it is desirable to wire a 13A socket from the load side of the cooker outlet, as well as connecting the hob to the cooker outlet as well.

Where the hob is sited in a different part of the kitchen from the oven, it is acceptable to simply plug the oven directly into the ring and use the cooker outlet for the hob. It is a requirement that the cooker switch is within 2 metres of the cooker. Where there is a distance between the hob and the oven, using a socket outlet for the oven is a good way of complying, and diversity would only be needed for the hob, which may reduce the cable size even further.

Diversity on complete installations

The same principle that was used for the cooker can pretty much be used for complete installations.

It would not be unusual for a dwelling to have the following circuits:

- One 32A cooker circuit
- Two 32A ring final circuits
- One 16A immersion heater circuit
- Two 6A lighting circuits

If added together, the total load could be as much as 124A. If we consider that the maximum rating of a single-phase domestic supply is 100A, with many of the supplies only 80A or even 60A, we can see that even a small installation will, in total, exceed the supply. If we were to install a 10kW electric shower we could add at least another 45A to this and the total demand would be getting close to double the rating of the supply.

This is, of course, where diversity helps us, as when we apply to the district network operator (DNO) for a supply they will need to be told the maximum demand taking into account diversity. This is because it is fairly obvious that not everything will be used together.

Diversity tables are available to help us (Table 5.1) but they are only intended as a guide and even if we use them the calculated demand in many installations will exceed the rating of the supply.

Table 5.1

Purpose of the final circuit to which diversity applies	Individual households including individual dwellings in a block	Small shops, stores, offices and small business premises
Lighting	66% of total demand	90% of total demand
Heating and power	100% of total load up to 10A and 50% of the remainder	100% of largest load and 75% of the remainder
Cooking appliances	The first 10A and then 30% of the remaining load. + 5A if a socket outlet is included	100% of the largest + 80% of 2nd largest then 60% of the remainder
Motors (not lift motors)		100% of largest motor + 80% of the 2nd largest and 60% of the remainder
Instantaneous water heaters (showers)	100% of the first two largest added and 25% of the remainder	100% of the first two largest added and 25% of the remainder
Thermostatic water heaters (immersion heater)	No diversity	No diversity
Floor heating	No diversity	No diversity
Storage and space heating	No diversity	No diversity
Socket outlet circuits, ring or radial (13A outlet circuits)	100% of largest demand + 40% of remainder	100% of largest demand + 50% of all others
Socket outlets other than those above and any fixed equipment not listed	100% of largest point used + 40% of all others	10% of largest used + 70% of all others

Using Table 5.1, let's look at applying diversity to the example we used earlier:

- One 32A cooker (diversity already applied to this)   = 32A
- One 32A ring final circuit at 100%   = 32A
- One 32A ring final circuit at 40%   = 12.8A
- One 16A immersion heater at 100%, 3kW heater   = 13A
- Two 6A lighting circuits at 66%   = <u>7.92A</u>
- Total   97.72 A

Note: where a fixed load such as an immersion heater is connected, the actual load can be used.

Using the diversity tables, this installation just about scrapes in below the rating of the supply fuse. If we were to add the shower that was mentioned earlier, the maximum demand will be much higher. In reality, the load will be far less than this.

Most domestic ring circuits will never be loaded up to their full capacity. The highest load in most houses will come from the kitchen when items of equipment such as the washing machine and dishwasher are used. Most other equipment in a kitchen are possibly high current but will be used for a short time only. Where a dwelling has central heating it is unlikely that the immersion heater will be used on a regular basis. Unless you have children like mine, it is unlikely that 66% of all lights will be on at the same time. Even if they were, it is extremely unlikely that each circuit will have nine 100W lamps, which is what would be required if the circuit load was to be 4A.

$$\frac{9 \times 100}{230} = 3.91A$$

When working out the maximum demand, common sense has to be used and we must remember that not everything will be on at the same time. In reality, most equipment will only be used intermittently.

It is fair to say that many installations have far more circuits and greater loads that the one used in our example. Ask around and see if you can find anyone whose supply fuse has operated due to overload. It is very unlikely that you will.

Diversity can be used to calculate maximum demand but realistic values must be used and a great deal of thought should be given to what equipment is likely to be used and when. You cannot simply keep adding circuits, because if you do, overloads can occur and the supply fuse may melt.

Chapter 6

Cable calculation

Basic cable calculation is a simple process providing you follow the correct sequence of steps.

Once the method of cable installation has been decided, it is time to look at selecting the correct type and size of cable that will be suitable for the job.

Of course, the type of cable used will depend on the installation, but the procedure for calculating the correct size is pretty much the same for all cables. It is really important, however, that the correct tables from the wiring regulations are used, particularly the tables from Appendix 4 of BS 7671.

As an example: a circuit to supply a 4kW wall mounted heater is to be installed in steel conduit using single core 70°C thermoplastic cables. The circuit is 36m long and the ambient temperature is not expected to rise above 30°C.

The circuit is to be protected by a BS EN 60898 Type B circuit breaker and the conduit will contain two other circuits.

The installation is a 230 volt single-phase TN-C-S supply with a measured Ze of 0.28Ω.

With the information available it is now possible to calculate the cable size a step at a time.

Step 1

Calculate the design current required for the load. This is referred to as I_b.

$$I_b = \frac{kW}{voltage}$$

$$\frac{4000}{230} = 17.39A$$

The design current I_b is 17.39A.

Now we have the design current we can carry out step 2, which is to select a protective device for the circuit.

Step 2

In the example we can see that we are to use a Type B to BS EN 60898 circuit breaker. The nearest rating of circuit breaker is going to be 20A.

Step 3

Step 3 is to look for any factors that may affect the temperature of the cables. These are known as rating factors.

Clearly, the temperature in which the cable is operating will be a factor. This is because if the cable heats up it will increase in resistance, and this will mean that its current carrying capacity will be reduced and the voltage drop will be increased.

The first factor to look at is the ambient temperature. Table 4B1 in Appendix 4 is the rating factor table and the symbol used to represent this factor is C_a.

Using the table we can see that C_a is 1.00 for a 70°C thermoplastic cable operating in an ambient air temperature of 30°C.

C_a is 1.00.

The next condition that will affect the cable is that it is installed in a conduit containing two other circuits. As we have no information other than that there are two circuits, we have to assume that the additional circuits are fully loaded and may be operating at a temperature of 70 (see note on operating conditions).

Table 4C1 is the table to be used for grouping and the symbol used to represent this is C_g.

From the evidence we have we can see that the circuit is enclosed in a conduit that will contain a total of three circuits. Using Table 4C1, this installation is item 1, the grouping factor for three circuits installed as item 1 the rating factor is 0.7.

Step 4

We now need to calculate the minimum size cable that will carry the current under the circuit installation conditions. The method that is used is to divide the rating of the protective device by the rating factors.

The information we need to use is:

- Rating of the protective device = 20A
- Rating factor for ambient temperature = 1
- Rating factor for cable grouping = 0.70

$$\frac{20}{(1 \times 0.70)} = 28.57A$$

This is the minimum rating of the cable that we can use under these conditions.

Step 5

Now we need to select a cable from the correct table in Appendix 4 of the wiring regulations. For single core 70°C thermoplastic insulated cables the table is 4D1A.

Column 4 is reference method B for cables enclosed in conduit or trunking fixed on to a wall. We now need to look down the column until we reach a figure of $\geq 28.57A$.

The nearest we can get is 32A, and if we now look across to the left we can see that in column one we need to use a 4mm² cable.

Now we have chosen a cable that can carry the load current we need to check that it meets all of the volt drop limitations this will be a maximum of 5% for this type of circuit.

Step 6

We now need to look at Table 4C1B. Column 3 is the mV/A/m and we can see that the volt drop for a 4mm² cable is 11 mV/A/m.

This is telling us that the voltage drop in our cable will be 11 thousandths of a volt for every amp and every metre. Our calculation will be just as it is shown – mV/A/m = 11 × 17.39 x = 6886 – but, of course, this is in millivolts and we need the answer in volts. To do this, we need to divide the answer by 1000. Rather than mess around doing the calculation in stages it is far easier to carry it out as one calculation.

Note that it is the design current that we use for volt drop.

$$\frac{11 \times 17.39 \times 36}{1000} = 6.88$$

This shows that our circuit will meet the volt drop requirements. In this instance the volt drop must be ≤ 11.5 volts as it is a power circuit with a limitation of 5%.

There is a shortcut that is often useful when calculating cables for volt drop. The information we need is the:

- Design current Ib = 17.39A
- Length of circuit = 36m
- Maximum volt drop for the circuit 5% = 11.5 volts

Once we have this information we can make life a little easier by transposing the original calculation of:

$$\frac{mV \times A \times L}{1000} = \text{volt drop}$$

Complete the bits we know for our circuit:

$$\frac{mV \times 17.39 \times 36}{1000} = 11.5$$

Now we can transpose to find the mV:

$$\frac{11.5 \times 1000}{17.39 \times 36} = mV$$

If we now complete the calculation we can see that the mV is 18.36mV/A/m.

All we need to do now is look at Table 4C1B and find a cable with a volt drop of less than 18.36mV/A/m. We can see that a 2.5mm² cable is less, and therefore it will comply with the volt drop requirements for the circuit. Now we need just to check that a 2.5mm² cable can carry the required current. Column 4 of Table 4C1A shows that the cable can only carry 24A, which will not be enough for us. All we need to do now is look down the column until we find a cable that can carry 28.57A and use that, as we know that the smaller cable of 2.5mm² is OK for volt drop.

There are other rating factors that could affect some circuits but will not affect the circuit that we are to install. Almost anything that will affect the temperature of the cable will have a rating factor.

In earlier editions of BS 7671 they were referred to as correction factors. Now they are known as rating factors but the initial letter used is still C.

The rating factors used are:

- C_a for ambient temperature
- C_g for grouping
- C_r for semi-enclosed fuses
- C_i for thermal insulation
- C_c for circuits buried in the ground
- C_d for the depth of burial

The tables required for use with these rating factors are all found in Appendix 4 with the exception of C_i. The table that can be used for this is Table 52.2, which can be found in Chapter 52 of BS 7671.

Earth fault loop impedance

Now that the cable size has been calculated, the next stage of the circuit design is to confirm that the circuit protection will operate within the required time in the event of an earth fault or a short circuit.

This again can be carried out in simple steps.

Step 1

We need to know the value of the external loop impedance Ze. This is given to us in the scenario as being 0.28Ω.

If the value is not available we will need to measure it using an earth fault loop impedance tester. To carry out this test the supply at the board being tested must be isolated.

Once isolated, disconnect the incoming earthing conductor and test between this conductor and the incoming supply. The measured value will be the Ze for the board being tested (Figure 6.1).

Step 2

Calculate the resistance of the circuit line and earth. This is known as $R_1 + R_2$ with R_1 being the resistance of the line conductor and R_2 the resistance of the circuit protective conductor.

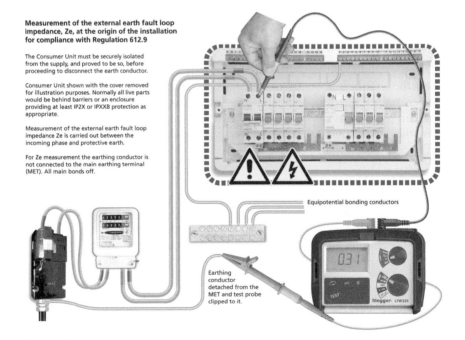

Measurement of the external earth fault loop impedance, Ze, at the origin of the installation for compliance with Regulation 612.9

The Consumer Unit must be securely isolated from the supply, and proved to be so, before proceeding to disconnect the earth conductor.

Consumer Unit shown with the cover removed for illustration purposes. Normally all live parts would be behind barriers or an enclosure providing at least IP2X or IPXXB protection as appropriate.

Measurement of the external earth fault loop impedance Ze is carried out between the incoming phase and protective earth.

For Ze measurement the earthing conductor is not connected to the main earthing terminal (MET). All main bonds off.

Equipotential bonding conductors

Earthing conductor detached from the MET and test probe clipped to it.

Megger. LTW325

Figure 6.1

The resistance values of copper conductors at a temperature of 20°C can be found in Table 6.1a below.

The cable being used for this circuit is 4mm² for both live conductors and cpc. From the table a 4mm² conductor has a resistance of 4.61mΩ/m. The circuit has a line conductor and a cpc, both of which are 36m long. The best method is to just add the resistance of the line and cpc together, which will give use a resistance of 9.22mΩ/m.

The calculation is to multiply the resistance per metre by the length, which will give us the total resistance of the cable.

$$\frac{9.22 \times 36}{1000} = 0.33\Omega$$

0.33Ω will be the resistance $(R_1 + R_2)$ of the cable at 20°C. When the cable is under load it may be that it will be running at a temperature of 70°C, so we will have to allow for the increase in conductor resistance by using a multiplier of 1.2 (this is explained in Chapter 3).

The resistance of $R_1 + R_2$ by the multiplier is:

$$0.33\Omega \times 1.2 = 0.396\Omega$$

0.396Ω will be the resistance of our circuit $R_1 + R_2$ at a temperature of 70°C.

Step 3

Next, we must add the $R_1 + R_2$ to the Ze as the calculation is usually set out as $Zs = Ze + (R_1 + R_2)$.

$$Zs = 0.23\Omega + 0.396\Omega = 0.626\Omega$$

Table 6.1a

Conductor cross-sectional area in mm²	Resistance of the conductor at 20°C mΩ/m	Twin and earth combinations mm²		$R_1 + R_2$ in mΩ/m
		Line	cpc	
1	18.10	1	1	36.20
1.5	12.10	1.5	1	30.20
2.5	7.41	2.5	1.5	19.51
4	4.61	4	1.5	16.71
6	3.08	6	2.5	10.49
10	1.83	10	4	6.44
16	1.83	16	6	4.23

The total earth fault loop impedance for our circuit when operating will be
0.626Ω.

Step 4

Now we need to compare the value against the values given for protective
devices in BS 7671, or if the table is not available it is a very simple exercise
to calculate the maximum permitted Zs for a circuit breaker to BS EN 60898.

The circuit breaker is a 20A Type B, which we know must operate within
0.1sec when five times its rating is passed through it.

$$20 \times 5A = 100A$$

To calculate resistance we use Ohm's law.
The voltage that the circuit is operating at is 230 volts.

$$\frac{V}{I}\Omega = \text{for our calculation } \frac{230}{100} = 2.3\Omega$$

We now have to adjust this value using the factor C_{min}, which is 0.95.

$$2.3\Omega \times 0.95 = 2.185\ \Omega$$

Providing our Zs value is $\leq 2.185\Omega$, the circuit is satisfactory.

Twin and earth 70°C thermoplastic PVC cable

When wiring circuits using 70°C thermoplastic, the process for cable selection is pretty much the same as the method shown in the scenario used in
the previous example. The differences will be the installation method and
the size of the cpc.

Generally, the cables are installed in the building structure, which may
contain insulation, and the cpc is smaller than the live conductors.

If we use the previous scenario we can see the difference: the rating factors will remain the same as will the circuit protection. This means that
the required rating of the cable will be the same – for our scenario it was
28.57A.

At this point of the calculation we need to consider the installation
method used to install the cable. Table 6.1b is used to identify the installation method for flat twin and earth. This table is used in conjunction with
Table 4D5A in BS 7671 to identify the current rating of a cable depending
on how it is to be installed.

There are four methods to choose from, shown in Table 6.1b.

Table 6.1b

Method 100	Flat twin and earth cables clipped to a joist, or touching the plasterboard ceiling but not covered by more than 100mm of thermal insulation
Method 101	Flat twin and earth cables clipped to a joist, or touching the plasterboard ceiling but covered by thermal insulation with a thickness of more than 100mm
Method 102	Flat twin and earth cables situated in a stud wall with the cable touching the inner wall surface or plasterboard ceiling
Method 103	Flat twin and earth cables situated in a stud wall not touching the inner wall surface and surrounded by thermal insulation

Let's assume that our cables are in a stud wall touching one surface of the wall – this will be method 102.

Now we need to look at Table 4D5A, which is specifically for use with flat twin and earth cables.

Column 4 is method 102; if we look down this column we can see that the nearest rating to 28.57A is 35A. Now look to the left of the table and we can see that the cable now needs to be a $6mm^2$ cable. Using this cable will make a difference to voltage drop and earth fault loop values. In this instance, the live conductors are a larger cross-sectional area, which means the voltage drop will be less due to the conductors having a lower resistance.

The earth fault loop impedance will need checking as the cpc in a twin and earth cable is usually smaller than the live conductors. A $6mm^2$ cable will normally have a $2.5mm^2$ cpc.

If we look at the cable resistance table we can see that the $6mm^2$ live conductors have a resistance of $3.08m\Omega/m$ and the $2.5mm^2$ cpc will have a resistance of $7.41m\Omega/m$. If we add them together we can see that the total resistance $(R_1 + R_2)$ per metre is $10.49m\Omega/m$.

If we now carry out the calculation as before using the new value:

$$\frac{10.49 \times 36 \times 1.2}{1000} = 0.45\Omega$$

Remember $Zs = Ze + (R_1 + R_2)$ or using our values $0.23\Omega + 0.45\Omega = 0.68\Omega$:

$$Zs = 0.6\Omega$$

This again, is much less than the maximum permitted for a 20A Type B BS EN 60898 circuit breaker, so the circuit will be OK.

There is one more regulation that we have to satisfy. Table 54.7 BS 7671 shows us that the minimum size of a cpc in any circuit with conductors

≤ 16mm² must not be smaller than the live conductors, unless the conductor size is calculated to ensure that it satisfies thermal constraints. In other words, we need to check that the cable will not be destroyed during the time it takes for the fault protection to operate.

For this, we need to carry out a calculation known as the adiabatic equation. This is the calculation used to check that the size of cpc we are using is suitable, *not* to calculate the size of the cpc. The calculation is:

$$S = \frac{\sqrt{I^2 \times t}}{k}$$

To carry out this calculation the information required is:

I = the value of fault current that can flow through the circuit including the protective device. This can be calculated using the Zs for the circuit and dividing it into the circuit voltage.

$$\frac{230}{0.68} = 338.23A$$

I = 338.23A

T = the operating time for the device. As it is a circuit breaker this will be 0.1 seconds.

k = the factor for the conductor taking into account the material, resistivity and heat capacity. This value can be found in Table 43.1 or Tables 54.2 through to 5 in BS 7671 depending on the conductor material. As this is 70°C thermoplastic the value will be 115.

S = is the minimum cross-sectional area of the cable in mm².

Using the values we now have we can complete the calculation:

$$\frac{\sqrt{338.23^2 \times 0.1}}{115} = 0.93mm^2$$

As our cpc is 2.5mm² it will be suitable.

The result of the calculation does not mean that we can reduce the size of the cpc. If we did that it would alter the $R_1 + R_2$ values, which in turn would alter the fault current and possibly the disconnection time. Always remember that this calculation is just a check.

Chapter 7

Socket outlet circuits

Most circuits that we install here in the UK are of the radial type where the cable simply loops in and out of each point and stops at the last outlet. However, an exception is made for circuits supplying socket outlets.

Ring final circuits

The most common type of socket outlet circuit in the UK, particularly in a domestic environment, is known as a ring final circuit (Figure 7.1). In this type of circuit the cable starts at the consumer unit; it then loops from one outlet to another and then back to the consumer unit forming a complete ring of cable.

The benefit of this type of circuit is that smaller cross-sectional area cables can be used – most commonly this would be 2.5mm² live conductors with a 1.5mm cpc. Also, the circuit can accommodate a large number of outlets and voltage drop around the circuit is normally very low.

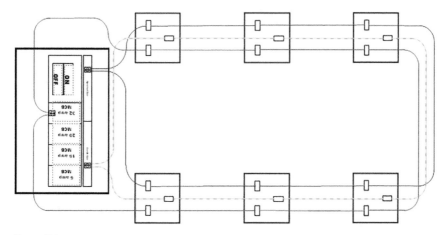

Figure 7.1

The rules for ring final circuits are as follows:

- Maximum area served is 100^{2}m.
- There is no limit to the number of outlets on a single circuit.
- A ring circuit can have as many socket outlets spurred from it as there are socket outlets on the ring.
- An unlimited number of fused connection units can be spurred from the ring.
- A spur cannot be taken from a spur.
- Maximum protective device size is 32A.

Wherever possible, it is always better to try and include all of the outlets within the ring but on occasion it is useful to spur off the ring, particularly if an additional socket is being added at a later date.

The advantage of a ring circuit is that socket outlets can be installed to supply a greater area while using a relatively small cable. However, it is very important that the circuit is wired correctly and any alterations to it are carried out carefully.

The disadvantages of using a ring circuit are that dangers can arise and not be obvious to users. An example would be where one of the conductors on the ring became open circuit (Figure 7.2). The circuit would operate normally, although the result would be that the cable is vulnerable to overload. The cable would only be capable of carrying a load of 20A but would be protected by a 32A device. Also, the value of the circuit Zs would increase and the circuit device may not operate in the event of a fault.

This is another good reason why periodic testing of installations is important.

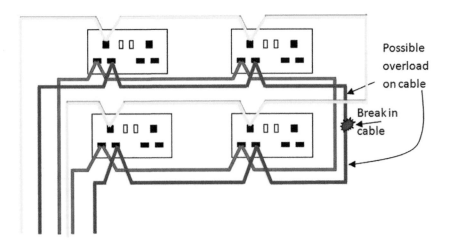

Possible overload on cable

Break in cable

Figure 7.2

Adding an outlet to an existing ring

As mentioned earlier, it is possible to add more sockets to a ring circuit if required but there are certain rules that have to be followed. When we add a socket to a ring it is known as a spur. We can spur as many double socket outlets to a ring as there are sockets on the ring. The socket on the ring can be a single or a twin socket.

When a socket has been spurred from the ring the cable supplying the socket is the spur and it is known as an unfused spur; the socket is *not* the spur. The spur can be taken from an existing socket outlet or a junction box can be cut into the circuit and the spur can be taken from that (Figure 7.3).

Under no circumstances can a spur be taken from a socket that is already a spur. A ring circuit is wired in 2.5mm² cable, which can safely carry a load of around 20A minimum; this depends on the installation method. The circuit is protected by a 30 or 32A protective device and the reason that this is permitted is because the circuit is wired as a ring, and the current can flow through the circuit along both legs of it. This, of course, provides a capacity of 40A, which will comply with our regulations as the cable rating is ≥ than the rating of the protective device.

When we spur a cable from the ring we must in most cases use the same size cable that the ring is wired in. When we spur a 2.5m² cable from the ring it is a single cable with a minimum current capacity of 20A. This is fine for one twin socket outlet as even if we load the socket outlet to its capacity it will not overload the cable, although the cable is rated at less than the protective device, it is permitted.

The reason for this is that any item of equipment that is plugged into the socket outlet will be fused in the plug top. This will prevent the equipment overloading and in turn the cable as a maximum load of only 26A will be

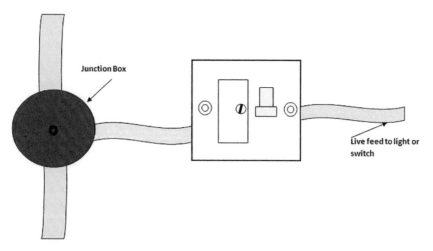

Figure 7.3

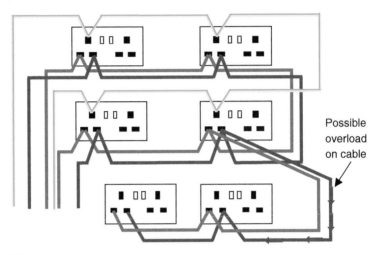

Possible
overload
on cable

Figure 7.4

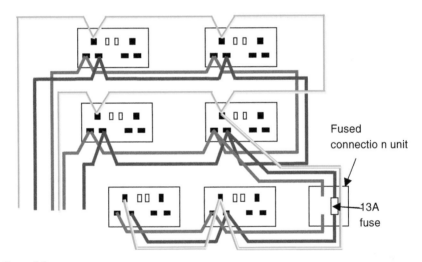

Fused
connectio n unit

13A
fuse

Figure 7.5

drawn. If a second outlet was spurred from the spurred outlet a maximum of 52A could be drawn, which of course would result in the overloading of the spur (Figure 7.4).

Where more than one outlet is required it is possible to spur from a socket outlet to a fused connection unit first. The cable in this instance would be called a fused spur as opposed to the unfused spur used for spurring a socket outlet. Once a fused connection unit has been spurred any number of socket outlets can be taken from the outgoing side of the spur as the 13A fuse in the connection unit will protect the cable (Figure 7.5).

Not all equipment has or even requires a 13A plug; very often, it is more convenient to permanently connect an item of equipment to a fused connection unit. It is permissible to spur as many fused connection units from a ring as required; there is no limit to the number providing a spur is not taken from a spur.

Radial circuits for socket outlets

Another method of installing socket outlets is by using a circuit called a radial; this method is preferred by many installers and designers as it is not as open to abuse as ring circuits.

A radial circuit consists of any number of socket outlets wired by looping a cable from the consumer unit to each socket in turn and terminating at the final outlet on the ring (Figure 7.6).

The circuit is usually wired in either 4mm^2 live conductors with a 1.5mm^2 cpc, or 2.5mm^2 live conductors with a 1.5mm^2 cpc.

Circuits wired in 4mm^2 can be protected by a device rated up to 32A and serve an area of up to 75m^2. Those wired in 2.5mm^2 can be protected by a device of up to 20A and serve an area of up to 50m^2.

The attraction for this type of circuit is that it cannot be compromised by breaks in cables and spurs being installed. In the event of the circuit conductors becoming disconnected or broken, all of the outlets after the break will not work, which is not the case with ring final circuits.

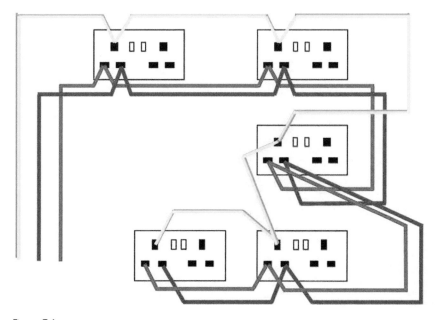

Figure 7.6

Spurs that are either fused or unfused can be added as and when required. Spurs can be taken from spurs with no risk of overload; this can be clearly seen in the radial diagram.

There is one point worth considering: where the circuit is used in areas where the load is likely to be high, such as a kitchen, it is advisable to use 4mm² live conductors. This is also good advice where ring final circuits are wired in areas with heavy loads.

Fused connection units

Fused connection units can be installed and used for various purposes. They can be switched or unswitched and they can also have an integral RCD. Where a fused connection unit is being installed on a ring circuit, there is no limit to the number of them that can be installed. However, the same rules apply for fused connection units on a ring. It is not permissible to spur from a unit that has already been spurred, unless of course the supply for them is taken from the load side of a spur (Figure 7.7).

Consideration should always be given to what the fused connection is going to be used for, as fused connection units must never be used to supply equipment that is rated at greater than 3kW. As with anything else in the

Figure 7.7

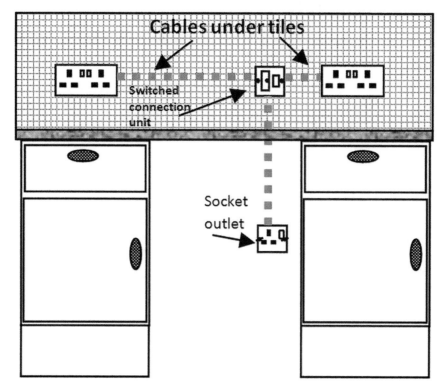

Figure 7.8

electrical world, it is better not to load a fused spur to its capacity, particularly where the load is to be on for long periods.

Where these situations cannot be avoided, it is always a good idea to use a good quality brand of connection unit. Any equipment that is used has to be to a BS or BS EN standard but this is the minimum standard required – for that reason some manufacturers are better than others.

Very often, fused connection units are used above kitchen worktops to provide isolation for socket outlets that are behind washing machines, dishwashers and similar pieces of equipment where the cord supplying the equipment needs to be hidden (Figure 7.8).

Where a fused connection unit is to be used for equipment such as alarm panels it is usual to use an unfused switch spur, which, of course, will prevent unintentional switching off of equipment.

If it is important to see at a glance whether or not the connection unit is on, a neon indicated spur can be used. Care must always be taken to ensure that the connection unit is connected correctly, with the supply and load being connected to the correct terminals.

Chapter 8

Safe isolation

Safe isolation is a very important skill and it is vital that it is carried out using the correct procedure. When carried out correctly, it eliminates as far as possible human error, which is the cause of most electric shocks. The Electricity at Work Regulations do not permit live working unless it is unreasonable in all circumstances not to work live, and even then all possible safety precautions have to be in place to make it reasonable to work live.

Of course, it is impossible for anyone involved in electrical work not to have to work live at some point, as live testing is a requirement and even safe isolation involves an element of live testing. This is acceptable but should be kept to a minimum and all possible safety precautions must be in place.

The correct equipment must be available and it must comply with all of the HSE safety recommendations. The HSE produce a document GS38, which sets out the safety requirements for the equipment used for measuring voltage current and resistance. This is a non-statutory document but compliance with it will ensure that the correct levels of safety with regards to testers, probes and test leads are achieved.

Equipment required

A voltage indicator such as a two pole voltage detector is the preferred device. This will have simple indicator lamps that will show the level of voltage present. This type of device is ideal as it has no switches and cannot be turned off or put on the wrong setting.

The probes must have the minimum of exposed metal tips, certainly no longer than 4mm (Figure 8.1).

Signs must be used to show that the circuit or system has been isolated intentionally (Figure 8.2).

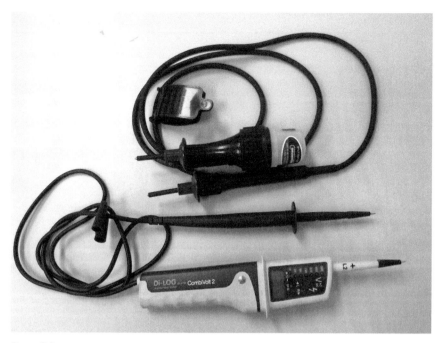

Figure 8.1

Figure 8.2

Figure 8.3

A range of lock offs must also be available. Unfortunately, there is no such thing as a universal lock off, which means that you have to have a good selection along with suitable padlocks (Figure 8.3).

Another useful piece of equipment is a proving unit. This is a device that can be used to check the correct operation of the voltage indicator (Figure 8.4).

Once you have made sure that you have all of the correct equipment in good working order and a suitable lock off for the type of device that you are going to lock off, it is safe to begin.

One important point is to make sure that anyone who is going to be affected by the loss of the supply is aware of what's going on.

Single-phase isolation

Step 1

Using the voltage indicator, check that the circuit to be isolated is live. This will also show that the voltage indicator is working. If the circuit is showing as being dead, then further checks are required to find out why – perhaps the voltage indicator or the circuit is faulty. Whatever it is, it has to be sorted out before the circuit can be isolated as you need to be sure that you are in control of the circuit that is being isolated.

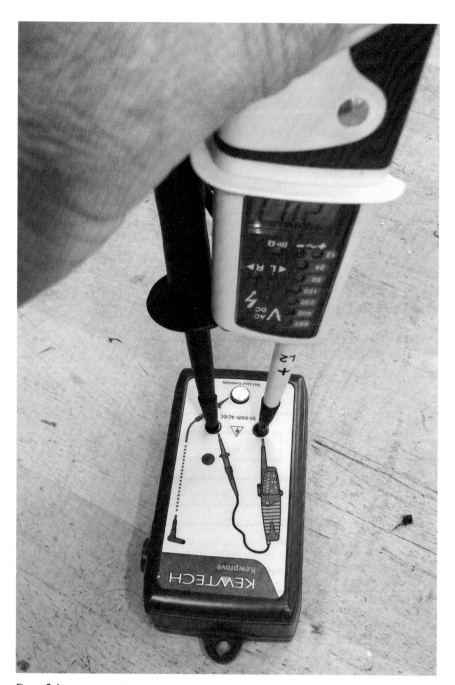

Figure 8.4

Step 2

Once you are sure the circuit is live you must identify the point of isolation and turn it off, lock it off and then put a sign on it to show that it has been locked off intentionally.

Step 3

Check that the circuit is dead by testing between the live conductors and then the live conductors and earth.

Step 4

Now check that the voltage indicator is still working by testing it on a known live supply, or if that is not possible the proving unit can be used.

Step 5

I always check again that the circuit is dead before I start work. This may be a belt and braces approach but it is better to be safe than sorry.

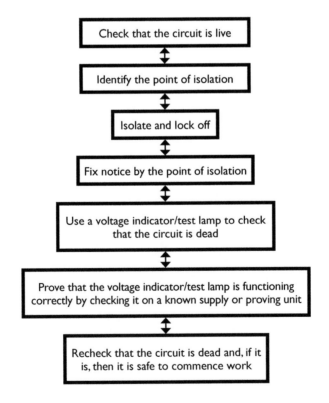

Figure 8.5

Where a piece of fixed equipment is supplied by a fused connection unit the same isolation procedure is used and the lock can be fitted through the fuse carrier (Figure 8.5).

Three-phase isolation

This requires the same equipment as single-phase isolation. The only additional equipment that may be needed are larger padlocks and lock off devices.

Step 1

Use the voltage indicator to check that the installation is live. You must test between all three lines, all three lines and neutral, all three lines and earth, then, last of all, neutral to earth. It often helps if you count the tests as you do them, as there are ten tests.

If, for some reason, the voltage indicator does not illuminate it is important to find out why. It could be that the voltage indicator is not working or is may be that the supply to the isolation point has been lost for some other reason. If that is the case you must find out why, as you need to ensure that the system is under your control.

Step 2

Switch off the device that is being used as isolation, lock off and put up a sign.

Step 3

Use the voltage indicator to carry out the ten steps described in step 1.

Step 4

Re-prove the voltage indicator on the proving device or a known live supply.

Step 5

Once again, I always repeat the ten-step test just to be sure.

Chapter 9

Isolation and switching

It is important that all electrical installations and all parts of the installation can be isolated safely.

The requirement of BS 7671 is that isolation must be possible at the origin of the installation, each circuit, every item of equipment and every motor.

One very important thing to remember about any device used for isolation is that it must be able to be locked in the open position or placed in a lockable enclosure/cupboard to prevent inadvertent or unauthorised closure.

Every means of isolation must have a facility allowing it to be locked off, and the on and off position must be clearly marked.

Isolators can be various devices and it is important that the function of each of these is understood.

A disconnector is a device used for isolation purposes only and should not be used for switching as it is not intended to be used under load.

A load switch is a device used for functional switching under load; some of these devices can be used for isolation as well. A lighting switch would be an example of a load switch but, of course, this cannot be used for isolation.

A switch disconnector is used to switch off under load and can also be used for isolation.

Circuit breakers are capable of making, carrying and breaking currents under normal conditions. They are also capable of breaking currents under fault conditions and in most cased are designed to be used as isolators.

Origin

For single and poly phased circuits a linked switch must be provided that will isolate the supply. This switch must also be able to interrupt the supply while it is on full load.

Where the isolator is for a single-phase supply that may be used by unskilled persons, it must be a double pole device.

Three-phase supplies can be treated slightly differently. Where the three-phase supply is part of a TT system it is important that the isolator interrupts all live conductors. This means all line conductors and the neutral.

Where the supply forms part of a TN-S or TN-C-S system only the line conductors need to be operated.

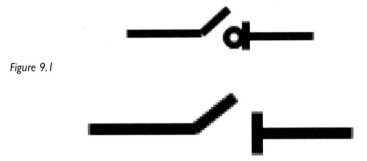

Figure 9.1

Figure 9.2

Any device used at the origin of the installation must be able to be switched while under load on load. This type of device will have a symbol that indicates that it is suitable for on load switching (Figure 9.1). If the device has a symbol shown in Figure 9.2, is should only be used for isolation off load.

Distribution boards and consumer units

Apart from the origin of the larger installations there may well be more than one distribution board or consumer unit; these boards must be able to be isolated without interrupting the supply to other boards or circuits. The main switch on these must switch all live conductors if the system is TT and all line conductors if the system is TN. Again, these devices must be capable of operating on load.

Circuit breakers

Every installation must be divided into circuits. This is to:

- Avoid danger and minimise inconvenience in the event of a fault.
- Allow for safe inspection, testing and maintenance.
- Take into account any dangers that may arise due to the failure of a single circuit.
- Reduce the possibility of nuisance tripping of RCDs due to the accumulation of protective conductor currents.
- Prevent unintended energisation of circuits intended to be isolated.

Isolation of circuits can be achieved by using circuit breakers and fuses. Where these devices are used in this situation, they will also be used to provide protection.

Motors

All fixed electric motors must have a means of isolation. All motors with a rating of 375W or greater must have not only a means of isolation but also be provided with control equipment that incorporates overload protection.

These are generally referred to as motor starters. The motor starter must also be able to prevent automatic restarting in the event of an overload or fault; this is often known as a no volt coil. Any isolation provided for a motor must also isolate the control gear.

Where the isolator cannot be installed in a position where it is obvious which motor it is being used to isolate, it must be clearly labelled.

As well as the need for circuits to be isolated, any items of electrical equipment supplied by the circuits must also be able to be switched on and off as well as isolated.

Although we need to switch equipment on and off, this switching on and off is often for different reasons and different terminology is used.

Functional switching

Functional switching is used where parts of a circuit or parts of the installation need to be able to be controlled independently of the rest of the installation.

A lighting point would be a good example.

There are various devices that may be used for functional switching; they must all be capable of switching under load. When installing switches for lighting it is always better to use a good quality switch, particularly when switching fluorescent fittings or extra low voltage luminaires supplied through a transformer. These types of light fittings may be inductive and when the switch is opened the arcing caused by the inductive loads may burn the switch contacts.

Clearly, functional switching will require the device to operate while under load, these types of devices could be:

- Light switches
- Cooker switches
- Passive infrared detectors
- Circuit breakers
- RCDs
- Isolating switches (not isolators); where isolating switches can be used for functional switching they must have the symbol ——⟋⊙— marked on them
- Where the device is marked with ——⟋ ⊢— this symbol indicates that it is suitable for isolation only; not switching
- Plug and socket outlets rated up to and including 32A
- Switched fused connection units
- Contactors

Although circuit breakers, RCDs and main installation switches are capable of switching off under load, they are not intended for regular/frequent load switching. In all cases it is preferable to use a switch that is intended and designed for frequent use.

Isolation

Although some devices used for isolation can be used to switch under load, the purpose of an isolator is to separate a part of the installation or sometimes the whole of an installation from every source of electrical energy.

Where the isolator is not a switched disconnector, it should not be used to switch loads. It is important to ensure any load is turned off using its functional switch before isolating the circuit.

Repeated use of an isolator to switch loads will result in pitting of the contacts, which in turn could result in a high resistance connection when the isolator is closed and under load. This, of course, will produce heat that could be harmful to the installation.

Plugs and sockets make ideal isolators, as do plug-in ceiling roses providing they are used to isolate the piece of equipment that you are working on.

Fused connection units are also very good for isolating small items of equipment; a padlock can be put through the fuse carrier (Figure 9.3).

Figure 9.3

When carrying out isolation on a TT or IT system it is important to remember that BS 7671 requires that all live conductors are isolated. This, of course, means all line conductors and the neutral conductor. On any TN-S or TN-C-S system, only the line conductors need to be isolated.

Switching off for mechanical maintenance

This is pretty much the same as isolation, although the device must be capable of switching the load. All must be labelled, and lockable in the off position.

The open position of the contacts must be clearly visible. A switch with the symbols O and I (Figure 9.4) would be suitable.

The devices used for this could be:

- Circuit breaker (for TT or IT systems this must be double pole for single phase and four pole for three phase and N, as all live conductors must be disconnected)
- Multipole switch
- Control switch operating a contactor or starter
- Plug and socket outlet up to and including 16A rating

Figure 9.4

Emergency switching

We tend to think that emergency switching is used just for switching equipment off. It is as well to remember that it can also be used for switching on.

The purpose of emergency switching is to control the supply in a manner that will remove any unexpected danger.

When installing an emergency stop system on TT and IT supply systems the same rules apply as for isolation and all live conductors must be separated from the supply.

Where emergency switching is required in a work area that may have several items of electrical equipment and socket outlets, the most practical method of providing emergency switching is to use a number of push buttons that break the control circuit to a contactor (Figure 9.5).

The contactor could be placed into the supply feeding a distribution board. The control a circuit would then be passed through a number of emergency stop buttons (Figure 9.6) located in accessible areas.

An emergency stop button once pushed must latch in the off position unless the stopping and re-energising of the button is under the control of the same person.

In a work environment where there are a few people working it is not unusual for the stop button to lock in the off position and the system to only

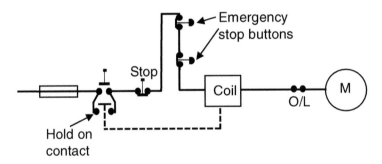

Figure 9.5

Figure 9.6

Figure 9.7

be able to be re-energised by the use of a key to unlock the button (Figure 9.7). This ensures that the system cannot be re-energised until the reason for the emergency stop has been found.

In most instances, the best device to use for emergency switching is a stop push button. Where these are used the button should be red with a contrasting background.

Firefighters' switch

Firefighters' switches are intended to be used by firefighters to switch off any external or internal lighting installations that exceed low voltage (1000 volts) (Figure 9.8).

Figure 9.8

Figure 9.9

Where the device is used to isolate exterior installations it should be placed outside of the building next to the equipment it is intended to isolate. If it is not possible to site the switch adjacent to the equipment, then a notice indicating the position must be fixed near the equipment.

For equipment sited inside the building, the switch should be installed at the main entrance. If this is not possible, then the position of the switch must be agreed with the local fire authority.

Unless otherwise agreed by the local fire authority, all firefighters' switches must be fixed so that they can be clearly seen and reasonably accessible to firefighters. They should be a maximum of 2.75m from the standing surface beneath the switch. All firefighters' switches must be coloured red and clearly marked with a label reading 'FIREFIGHTERS' SWITCH' (Figure 9.9). The size of the label must be a minimum of 150mm × 100mm with a minimum of 36 point lettering.

The switch must be clearly marked on and off, with the off position being at the top. There must be a mechanical device to prevent the switch being returned to the off position accidentally.

Chapter 10

Periodic inspection volt drop

Calculation of volt drop when carrying out a periodic inspection and test

As we know, it is now a requirement of BS 7671 that volt drop is verified for each circuit (reg. 612.14).

Voltage drop can be verified by measuring the voltage at the origin of the supply and then again at the furthest point on the circuit. Of course, this requires the circuit to be loaded up to its maximum.

This is simple process when it is a fixed load such as an immersion heater or a room heater, but it can be more difficult if we are dealing with a socket outlet, cooker or lighting circuit.

Calculation is by far a much simpler process, which can be done using dead test values. To carry out a periodic inspection we must have available the $R_1 + R_2$ value for each circuit.

As a check we can simply multiply the $R_1 + R_2$ value by either the load current if it is known, or the rating of the circuit protective device. This must then be multiplied by a factor of 1.2, which is to correct for the rise in temperature when the circuit is under load.

Often, the R_2 will be a conductor with a smaller CSA than the live conductors. This, of course, will result in the calculation showing a higher voltage drop than there would be in reality, which doesn't matter providing the volt drop calculated is less than the permitted value. Remember, this is only a check!

Where this calculation gives a value greater than that permitted it may be that if the $R_1 + R_N$ value was to be used in the calculation the volt drop would possibly comply. Of course, many test certificates will not show the value for R_N but don't despair, as a simple calculation will give you all of the information you need.

As an example, let's take a circuit that is wired using a 2.5mm²/1.5mm² twin and earth cable that has an $R_1 + R_2$ value of 0.6Ω. The circuit is protected by a 20A device.

Using the $R_1 + R_2$ value of 0.6Ω, we can now calculate the voltage drop for the circuit.

$$0.6 \times 1.2 \times 20 = 14.4 \text{ volts}$$

Clearly this is far too high. However, if we use $R_1 + R_N$ we may end up with an acceptable result. The problem is we do not have R_N but we can find out what it is by using a very simple calculation.

$$\frac{CSA_{line}}{CSA_{line} + CSA_{cpc}} \times (R_1 + R_2) - R_2$$

To put figures to this:

$$\frac{2.5}{2.5 + 1.5} = 0.625 \times 0.6 = 0.375\Omega$$

0.375Ω is the resistance of R_2. Therefore if we subtract this value from the $R_1 + R_2$ value, we will have the value of R_1.

$$0.6\Omega - 0.375\Omega = 0.225\Omega$$

The resistance of the line conductor is 0.225Ω.
If we now double this value we will have the value of $R_1 + R_N$.

$$0.225 + 0.225 = 0.45\Omega$$

Now we can carry out the voltage drop calculation using 0.45Ω as the resistance value.

$$0.45 \times 20 \times 1.2 = 10.8 \text{ volts}$$

This shows the volt drop to be OK.
As you can see, this is a simple calculation, which once it has been performed a couple of times will become second nature. It is also much easier than trying to measure the volt drop by applying loads to circuits.
An added bonus is that it can be carried out in the comfort of your van.
If you just want to use the $R_1 + R_2$ values, then Table 10.1 will indicate whether or not the readings are acceptable.
When using the table it is important to remember that it should only be used when the $R_1 + R_2$ values have been measured accurately with no parallel paths present. Where parallel paths are present, lower inaccurate voltage drop values will be shown.

Table 10.1

Conductor size mm^2	Measured $R_1 + R_2$	Max load current (amps)	Volt drop	Max length (m)
10mm^2–10mm^2	0.151	63	<11.5	41.43
10mm^2–10mm^2	0.09	63	<6.9	24.87
10mm^2–10mm^2	0.191	50	<11.5	52
10mm^2–10mm^2	0.114	50	<6.9	31.33
10mm^2–10mm^2	0.238	40	<11.5	65.25
10mm^2–10mm^2	0.143	40	<6.9	39.3
10mm^2–10mm^2	0.297	32	<11.5	81.3
10mm^2–10mm^2	0.178	32	<6.9	48.97
10mm^2–10mm^2	0.475	20	<11.5	130.5
10mm^2–10mm^2	0.286	20	<6.9	78.61
10mm^2–10mm^2	0.599	16	<11.5	163.5
10mm^2–10mm^2	0.360	16	<6.9	98.2
10mm^2–4mm^2	0.266	63	<11.5	41.43
10mm^2–4mm^2	0.160	63	<6.9	24.87
10mm^2–4mm^2	0.33	50	<11.5	52
10mm^2–4mm^2	0.201	50	<6.9	31.33
10mm^2–4mm^2	6.5	40	<11.5	65.25
10mm^2–4mm^2	0.25	40	<6.9	39.3
10mm^2–4mm^2	0.523	32	<11.5	81.3
10mm^2–4mm^2	0.315	32	<6.9	48.97
10mm^2–4mm^2	0.840	20	<11.5	130.5
10mm^2–4mm^2	0.506	20	<6.9	78.61
10mm^2–4mm^2	1.05	16	<11.5	163.5
10mm^2–4mm^2	0.632	16	<6.9	98.2
6mm^2–6mm^2	0.239	40	<11.5	38.79
6mm^2–6mm^2	0.143	40	<6.9	23.36
6mm^2–6mm^2	0.298	32	<11.5	48.48
6mm^2–6mm^2	0.178	32	<6.9	29.5
6mm^2–6mm^2	0.478	20	<11.5	77.58
6mm^2–6mm^2	0.286	20	<6.9	46.72
6mm^2–6mm^2	0.597	16	<11.5	96.97
6mm^2–6mm^2	0.357	16	<6.9	58.4
6mm^2–6mm^2	0.735	13	<11.5	119.35
6mm^2–6mm^2	0.44	13	<6.9	71.7
6mm^2–6mm^2	0.956	10	<11.5	155.16
6mm^2–6mm^2	0.572	10	<6.9	93.44

(Continued)

Table 10.1 (Continued)

Conductor size mm²	Measured $R_1 + R_2$	Max load current (amps)	Volt drop	Max length (m)
6mm²–2.5mm²	0.406	40	<11.5	38.79
6mm²–2.5mm²	0.245	40	<6.9	23.36
6mm²–2.5mm²	0.508	32	<11.5	48.48
6mm²–2.5mm²	0.309	32	<6.9	29.5
6mm²–2.5mm²	0.81	20	<11.5	77.58
6mm²–2.5mm²	0.49	20	<6.9	46.72
6mm²–2.5mm²	1.01	16	<11.5	96.97
6mm²–2.5mm²	0.612	16	<6.9	58.4
6mm²–2.5mm²	1.25	13	<11.5	119.35
6mm²–2.5mm²	0.752	13	<6.9	71.7
6mm²–2.5mm²	1.627	10	<11.5	155.16
6mm²–2.5mm²	0.98	10	<6.9	93.44
4mm²–4mm²	0.299	32	<11.5	32.4
4mm²–4mm²	0.18	32	<6.9	19.5
4mm²–4mm²	0.478	20	<11.5	51.84
4mm²–4mm²	0.285	20	<6.9	31
4mm²–4mm²	0.597	16	<11.5	64.8
4mm²–4mm²	0.359	16	<6.9	39
4mm²–4mm²	0.734	13	<11.5	79.7
4mm²–4mm²	0.442	13	<6.9	48.01
4mm²–4mm²	0.958	10	<11.5	104
4mm²–4mm²	0.577	10	<6.9	62.65
4mm²–4mm²	1.59	6	<11.5	172
4mm²–4mm²	0.955	6	<6.9	103
4mm²–1.5mm²	0.541	32	<11.5	32.4
4mm²–1.5mm²	0.325	32	<6.9	19.5
4mm²–1.5mm²	0.866	20	<11.5	51.84
4mm²–1.5mm²	0.51	20	<6.9	31
4mm²–1.5mm²	1.08	16	<11.5	64.8
4mm²–1.5mm²	0.65	16	<6.9	39
4mm²–1.5mm²	1.33	13	<11.5	79.7
4mm²–1.5mm²	0.247	13	<6.9	48.1
4mm²–1.5mm²	1.73	10	<11.5	104
4mm²–1.5mm²	1.04	10	<6.9	62.65
4mm²–1.5mm²	2.87	6	<11.5	172
4mm²–1.5mm²	1.72	6	<6.9	103

Table 10.1 (Continued)

Conductor size mm²	Measured $R_1 + R_2$	Max load current (amps)	Volt drop	Max length (m)
2.5mm²–2.5mm²	0.354Ω	27	<11.5	23.88
2.5mm²–2.5mm²	0.212	27	<6.9	14.3
2.5mm²–2.5mm²	0.48	20	<11.5	32.38
2.5mm²–2.5mm²	0.287	20	<6.9	19.5
2.5mm²–2.5mm²	0.59	16	<11.5	39.81
2.5mm²–2.5mm²	0.355	16	<6.9	23.97
2.5mm²–2.5mm²	0.737	13	<11.5	49.70
2.5mm²–2.5mm²	0.44	13	<6.9	29.93
2.5mm²–1.5mm²	0.46Ω	27	<11.5	23.88
2.5mm²–1.5mm²	0.277	27	<6.9	14.3
2.5mm²–1.5mm²	0.755	20	<11.5	32.38
2.5mm²–1.5mm²	0.16	20	<6.9	19.5
2.5mm²–1.5mm²	0.77	16	<11.5	39.81
2.5mm²–1.5mm²	0.2	16	<6.9	23.97
2.5mm²–1.5mm²	0.96	13	<11.5	49.70
2.5mm²–1.5mm²	0.26	13	<6.9	29.93
1.5mm²–1.5mm²	0.479	20	<11.5	16.49
1.5mm²–1.5mm²	0.288	20	<6.9	9.9
1.5mm²–1.5mm²	0.598	16	<11.5	20.6
1.5mm²–1.5mm²	0.36	16	<6.9	12.4
1.5mm²–1.5mm²	0.736	13	<11.5	25.3
1.5mm²–1.5mm²	0.44	13	<6.9	15.24
1.5mm²–1.5mm²	0.96	10	<11.5	32.89
1.5mm²–1.5mm²	0.57	10	<6.9	19.81
1.5mm²–1.5mm²	1.59	6	<11.5	54.59
1.5mm²–1.5mm²	0.954	6	<6.9	32.88
1.5mm²–1.0mm²	0.497	20	<11.5	16.49
1.5mm²–1.0mm²	0.29	20	<6.9	9.9
1.5mm²–1.0mm²	0.622	16	<11.5	20.6
1.5mm²–1.0mm²	0.374	16	<6.9	12.4
1.5mm²–1.0mm²	0.764	13	<11.5	25.3
1.5mm²–1.0mm²	0.46	13	<6.9	15.24
1.5mm²–1.0mm²	0.99	10	<11.5	32.89
1.5mm²–1.0mm²	0.598	10	<6.9	19.81
1.5mm²–1.0mm²	1.64	6	<11.5	54.59
1.5mm²–1.0mm²	0.99	6	<6.9	32.88

(Continued)

Table 10.1 (Continued)

Conductor size mm²	Measured $R_1 + R_2$	Max load current (amps)	Volt drop	Max length (m)
1.0mm²–1.0mm²	0.598	16	<11.5	16.51
1.0mm²–1.0mm²	0.36	16	<6.9	9.94
1.0mm²–1.0mm²	0.95	10	<11.5	26.41
1.0mm²–1.0mm²	0.57	10	<6.9	15.9
1.0mm²–1.0mm²	1.59	6	<11.5	44.26
1.0mm²–1.0mm²	0.957	6	<6.9	26.66

Chapter 11

Installing an electric shower

The first thing to remember when installing an electric shower is that if it is in a dwelling, it will be notifiable under Part P of the building regulations and an electrical installation certificate must be completed.

There are some important things that must be checked first.

The electric shower must have a mains water supply. It is unlikely that it will work correctly off stored water system because the water pressure or flow rate will not be high enough. Most electric showers will require a minimum pressure of 1 bar and a flow rate of 8 litres per minute, but it is a good idea just to check the data for the type and model of shower that you intend to fit. Data sheets and instruction manuals can usually be found on the Internet.

Once you have checked that the water supply is suitable, the next step is to check that the electricity supply is suitable and capable of carrying the additional load. Again, the current rating of the shower can be found on the shower data sheet. If the rating is given in kilowatts (kW) a simple calculation will provide you with the current rating. Let's say that the shower is going to be rated at 9.5kW. The current rating can be found by:

$$\frac{\text{power rating}}{\text{volts}} \text{ (in this case } \frac{9500}{230} = 41.3 \text{ A)}$$

Having found the current rating you now need to look at the consumer unit to see if there is a spare way that will be required to accommodate the new circuit. You will also need to check that the supply and existing installation is suitable for the additional load. Items that need to be checked are:

The rating of the supply fuse and tails

If the supply fuse is 80A or 100A, there would not normally be a problem in a domestic installation unless, of course, there were a lot of heavy loads connected that will be used at the same time as the new shower. Where the supply fuse is 60A more consideration should be given to the other loads connected to the installation. As an example, are they likely to be used at the same time as the shower? If they are, will the total load being used exceed the rating of the supply? You have to remember that the additional load while it's operating is two-thirds of the installations capacity.

It's not a bad idea to ask the occupier if they have ever had problems with the supply fuse. If you have any concerns it may be a good idea to speak to the supply company and ask them to change the service head for a higher current rating. This will not normally present a problem.

Condition of the consumer unit

It really doesn't matter if the consumer unit and protective devices are old as long as they are in good condition – rewirable fuses are perfectly accept-able to use. You must remember, though, that the new circuit must comply with the current edition of BS 7671. This, of course, means that the shower circuit will need to be RCD protected. If there are no spare ways, then the choices are to upgrade the consumer unit or install a Henley block in the tails and fit a second small unit for the shower (Figure 11.1).

Earthing and bonding

When a new circuit is to be installed, the earthing and bonding must be compliant with the current edition of BS 7671. The size of the earthing and bonding conductors will be dependent on the rating of the supply. How-ever, if it has to be upgraded, then it is just as easy to install the maximum required sizes. For a 100A TN supply the earthing conductor will be 16mm^2 and the main bonding conductors will be 10mm^2. For a TT system, the earthing conductor can be 4mm^2 in some cases.

Now that you have confirmed the suitability of the water and electricity sup-ply it is time to consider the actual installation of the supplies to the shower.

Water

As described earlier, the water supply must be of a sufficient pressure and flow rate, which will usually require connection to the main water supply. If the system is one that uses tanks to store cold water, usually in the roof, it is usually quite simple to find the parts of the system that are fed by the mains. Hot taps will not be fed by the mains so it is not worth considering those. Main supply water is going to be found at the cold tap in the kitchen and the supplies to the tanks in the roof, possibly in other cold taps around the building and often the supplies to WC cisterns. The easiest way to find out is to turn off the water at the main stop cock and then go around the house turning on taps and flushing toilets. If there is no water at the tap, or the cistern of the toilet does not refill, you will have found a part of the

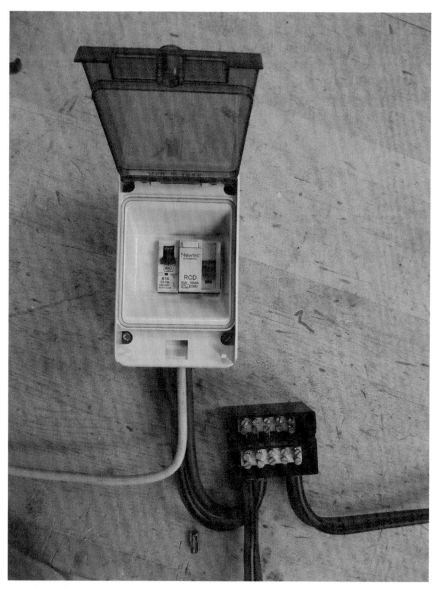

Figure 11.1

system that is connected to the main supply. It is normally a good idea to turn on the kitchen cold tap first and let it run to drain out any water that may be in the pipes.

It's fairly obvious that you need to decide on which is the simplest route to run the water pipe to the shower. In most cases, the best place to find a main supply that is easy to get to the shower is to connect to the pipe work supplying the tank in the roof. The pipe supplying the water to the ball valve is the one to look for. When installing the water supply to the shower, remember to fit an isolating valve that is reasonably accessible so that once the pipe has been connected to the supply it can be isolated, and the water to the rest of the building can be turned on.

If the water system in the house is an unvented system with no cold water storage all of the cold taps and cisterns will be connected to the main water supply and will be suitable to connect to. You will need to make sure that it is a pipe supplying cold water, though.

Electricity

With regard to the electrical installation, it is important to remember that all circuits supplying equipment in a bathroom must have additional protection. This requires that the circuits are protected by an RCD with a maximum trip rating of 30mA.

In the calculation we used a 9.5kW shower that had a current rating of 41.3A. The next step is to select a protective device. Assuming the protection is a circuit breaker, the nearest rating above 41.3A will normally be 50A and it would need to be a Type B.

There is not really a need to get into cable calculations when fitting a shower, as the manufacturer's instructions will dictate the size of the cable that you should use. It may be that you can calculate a smaller size but you should remember that manufacturers' instructions must always be followed. Another thing to remember is that because the shower circuit cannot overload there is no need to provide overload protection, although fault protection must be provided.

In other words, the circuit conductors can be rated at a lower current than the protective device because the shower is a fixed resistive load that will either operate up to its full load, or not operate at all. Of course, it could short-circuit if, for example, the element were to corrode, but fault protection will prevent any damage being done to the circuit.

Having decided on the cable size and current rating of the circuit protection, the cable can be installed. Of course, the method of installation depends on the type of building and the route that the cable will follow. Where possible, it is better to avoid any thermal insulation or indeed anything that may get hot; never be tempted to install cables in notches alongside hot pipes.

Where thermal insulation cannot be avoided it will not present too much of a problem, as the nature of a shower means that it is not going to be drawing current for more than a few minutes in most cases. However, every installation must be considered on its own merits and sometimes judgements need to be made.

All pieces of electrical equipment must have a suitable method of isolation and any fixed equipment installed in a room containing a bath or a shower will require a double pole switch as a method of isolation. The switch can be inside or outside of the room and it can be a plate switch-mounted on a wall or a cord switch on the ceiling. The choice of switch often depends on where it is to be installed.

Rooms containing a bath or a shower have particular zones (Figure 11.2).

The switch for a shower must be outside of zone 2, although if a cord switch is used the cord is permitted to hang down into zone 1. Wherever possible, I prefer to use a wall-mounted switch as sometimes the large cables required for electric showers can make the connections in a cord switch difficult.

When a cord switch is used it is always a good idea to use one with an indicator lamp, which will enable the user to see if the device is switched on or off. This is not normally as important with a wall-mounted switch as the on and off functions are marked and much easier to see.

Before installing a shower it is important that consideration is given to the positioning of the shower. Ideally, the shower should be able to be turned on

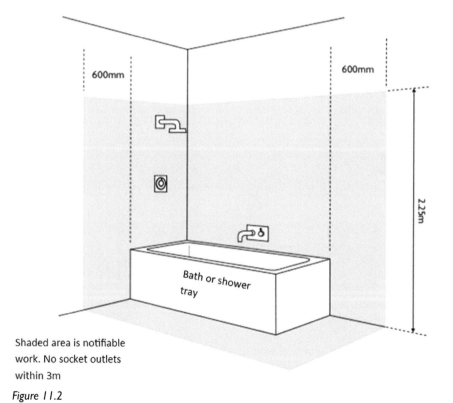

Shaded area is notifiable
work. No socket outlets
within 3m

Figure 11.2

without the person using it getting wet before they want to. Remember, the water from an electric shower will be cold when first switched on.

When the shower installation has been completed, the first check should be for any water leaks. Remember, the water is at mains pressure and generally any leaks will appear quite quickly. Clearly, the water installation must be tested before any joints are covered up.

Once you are sure the water is OK, then the electrical initial verification should take place, and an electrical installation certificate must be completed and given to the local building control.

Chapter 12

Two-way and intermediate switching

There are many situations that require a lighting point to be controlled by more than one switch. An example of this would be in a dwelling that has stairs and the landing/hall light needs to be switched at the bottom and the top of the stairs. This type of switch circuit is known as two-way switching.

The circuit diagram shown shows how the circuit works (Figure 12.1).

A junction box system used in this type of circuit would be fine and can be connected just as the circuit shows. In the circuit diagram it can be seen that the permanent live is fed into the common of one switch, and the switched return is taken from the common of the other switch (Figure 12.2). This type of wiring is fine for a junction box system or a conduit system where single core cables are used, but it is not very practical for use in a three-plate system.

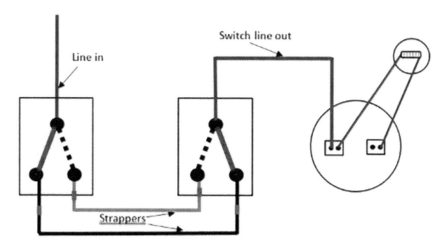

Figure 12.1

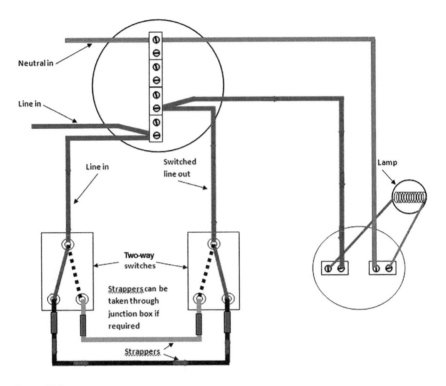

Figure 12.2

Where the system is a three-plate loop in system the most common type of two-way wiring circuit used is an extended two-way circuit (Figure 12.3). In this type of circuit all that is required is that three-core cable is run from the one-way switching position to the second switch position.

The standard colours of the cores in a three-core cable are brown, black and grey. As they are all to be used as line conductors, they must have a small piece of brown sleeving placed over them to identify them. When

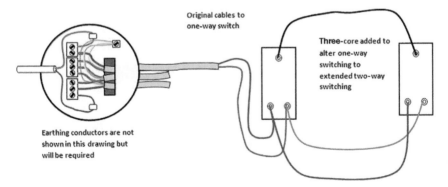

Figure 12.3

all of the cables have been installed and the outer sheathing removed, one switch should have two brown conductors and one each of blue, black and grey. The second switch will have one each of brown, black and grey. Both switches should have two bare cpcs, which will need to be covered completely with green and yellow earth sleeving.

When making the connections, the colours used are for identification only; the method shown in Figure 12.3 is the one that I would use. Providing the conductors are connected in the terminals as shown the colours are not important.

The switches required will be two-way switches and will have three terminals, which will be identified as common, L1 and L2 (Figure 12.4).

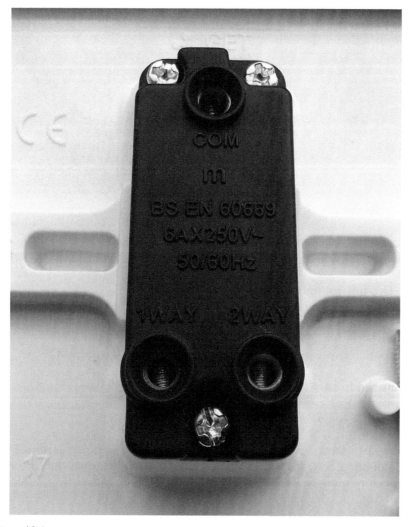

Figure 12.4

Figure 12.5

At switch one with five line conductors, connect (Figure 12.5):

- Two browns into terminal L1
- Black and blue into terminal L2
- Grey into common

At switch two with three line conductors, connect (Figure 12.6):

- Brown to L1
- Black to L2
- Grey to common

The cpcs must be joined and terminated in the earthing terminal of the mounting box. If there is not a terminal provided, then the earthing conductors must be connected into a suitable connector such as a block connector.

Figure 12.6

Intermediate switching

'Intermediate switching' is the term given to a system where the same light or lights can be switched on and off from a number of positions (Figure 12.7).

As with the two-way system, the circuit is fine for a junction box system but not practical for use with a three-plate system. A much simpler method is to use the same method as an extended two-way system and add switches as required by looping the three-core cable in and out of the additional switches.

At switch one there will be a switch line and return from the light and a three-core, which will go to switch two.

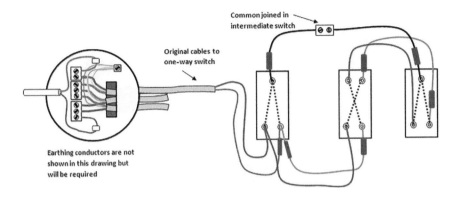

Common joined in
intermediate switch

Original cables to
one-way switch

Earthing conductors are not
shown in this drawing but
will be required

Figure 12.7

Switch two will have one three-core in and one three-core out. This will be the same for all of the switches apart from the final switch, which will only have one three-core.

The first and last switch will be two-way switches with three terminals. The other switches will be intermediate switches that will have four terminals marked as illustrated in Figure 12.8. Be very careful, as a double pole switch will look very similar.

The connections at the switches are as follows (*all conductors to have brown identification*).

Switch one

- Two brown conductors into terminal L1
- Black and blue conductors into terminal L2
- Grey conductor into the common

At switch two and all other intermediate switches

- Grey conductors to be joined and terminated into a connector.
- Brown and black conductor of one cable terminated in L1 and one in L2 on top or bottom of the switch.
- Brown and black conductor of the other cable terminated to the remaining L1 and L2.

It is very important that the conductors of each cable are terminated at the same end of the switch (*top or bottom*).

Figure 12.8

At the final switch

- Grey terminated into the common terminal
- Brown terminated into L1
- Black terminated into L2

Providing all of the connections are made correctly, the light/s will be able to be controlled from any switch position.

Chapter 13

Fault finding in central heating

There is one important rule with fault finding on new central heating systems that are just being commissioned, and that is to always check your wiring first, before looking for faulty components. Of course, if the system has been working correctly for a while, then it can be assumed that the wiring is correct, unless, of course, someone has been messing around with it.

Room thermostats

Although room thermostats can be seen as simple adjustable switches that just turn on the central heating when the temperature drops below the set level, they are a little more complicated than that.

Many room thermostats have what is known as an anticipator fitted (Figure 13.1), which is a small heater that operates when the system is calling for heat. The purpose of the anticipator is to prevent a large temperature difference between the on and off action of the switch.

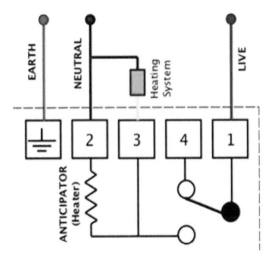

Honeywell room stat connection

Figure 13.1

The anticipator is a small heater that only turns on when the switch is closed; the switch is usually a bi-metal strip, which of course bends with the change of heat. This type of thermostat requires a neutral at the thermostat position and great care must be taken when connections are made; if the line and switch return are connected around the wrong way the device will still operate and the user may not notice the difference. There will be a large dwell between on and off temperatures and the system will use more energy.

Problems often occur when room stats are removed and a note of the connections is not made. When this happens it is important that the line, switched line and neutral are identified correctly.

Method

1   With the system isolated, separate all of the ends at the room thermostat.
2   Now turn on the heating system but turn off the central heating and hot water at the programmer. Using a voltage indicator test between each conductor to earth; there should be no voltage.
3   Now turn on the central heating at the programmer. Using a voltage indicator, test each conductor to earth – the one that lights up is the incoming line.
4   With the central heating still on, test between the known incoming line and the other two conductors. The one that indicates 230 volts will be the neutral.
5   Now isolate the supply to the heating system. Use the isolation procedure to prove it is dead and then connect the room thermostat, ensuring that the line and switch return are to the correct terminals (Figure 13.2).

Figure 13.2

It should also be noted that some electronic room stats do not require a neutral conductor.

Programmable room stats

These are devices that work as a room stat and also as a timing device. The benefits that are associated with using a programmable room stat are that they are usually wireless, and they allow the heating to not only come on at various times of the day but also at different temperatures. The biggest problem with using this type of room stat is that they can be difficult to set up, and most of the problems associated with them are due to the user fiddling with the settings, and not understanding how to set it back up.

Cylinder thermostats

Most cylinder stats are quite simple devices with three contacts that are marked as C (common), NO (normally open) and NC (normally closed) (Figure 13.3). These devices rarely go wrong and the correct operation of them is easy to test.

Obviously, the circuit needs to be isolated first. Then disconnect the cable from the terminal marked C and test for continuity between C and NO and then C and NC. If the cylinder is up to temperature the resistance readings will be low resistance between C and NO (closed circuit) and high resistance between C and NC (open circuit), as the stat will have operated to switch off

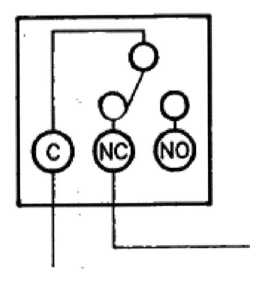

Figure 13.3

Figure 13.4

the heating. If you turn the stat up the reading between C and NC will be a closed circuit and between C and NO will show as an open circuit.

Where a cylinder stat has been connected to an unvented hot water cylinder the wiring will also pass through a thermal cut out (Figure 13.4). This is required as a secondary safety device and provides a second level of protection. Most immersion heaters are now fitted with their own thermal cut outs; if you have one without a thermal cut out you can wire the immersion through the cut out that is part of the cylinder.

Unvented cylinder

Where unvented cylinders are used on a wet heating system there must be at least three levels of protection. These would consist of a cylinder stat to control the water temperature, a thermal cut out to prevent the water boiling if the cylinder stat malfunctioned, and a motorised valve that is spring loaded to close when the supply to it is cut off due to high temperature (Figure 13.5).

These safety devices are very important and should never be ignored. Unvented cylinders work at mains pressure and when the water heats up it expands. The cylinder is basically a sealed container, which, of course, will have no room for expansion. For this reason, unvented cylinders must have expansion vessels fitted so that when the water expands it will be safe.

Figure 13.5

A major problem would be if the water were to boil, as when water is converted to steam at atmospheric pressure it expands by a ratio of 1:1700. While the water and steam are contained in a pressurised cylinder they will not present a problem other than being really hot. The problem arises if the cylinder ruptures under pressure, because as the boiling water escapes into atmospheric pressure it expands 1700 times. This is like a bomb going off and will literally destroy buildings and kill.

This is why there are three levels of protection. The first is the cylinder stat, which just controls the water temperature; the second, which is the overheat limit stat, is in place in case the cylinder stat does not operate correctly – the overheat stat must trip out and only be able to be reset manually. The third level would be installed where the cylinder is heated from a wet heating system; a spring return two-port valve should be fitted to the secondary heating coil with the valve wired through the overheat limit stat. Once the limit stat trips, the supply that is holding the valve open will cut off and the valve will close, preventing any more hot water passing through the cylinder. There are other temperature relief valves fitted but they are non-electrical.

Heating circulation pump

Most modern heating fluid-filled heating systems rely on a pump that is used to circulate the heated fluid around the system. Heating pumps are very reliable but, of course, as with any mechanical/electrical component they occasionally stop working. Often, it is a fault of another component but sometimes it is the pump itself.

Common problems

Where the pump is used for heating only and is not needed during the summer months it sometimes seizes up, and when the heating is turned on the boiler may work but the heating system does not because the pump is not circulating any heating fluid round the system.

The first step is to check that the pump is not running. Just because the heating is not on does not mean that it is the pump at fault. It could be that a motorised valve has not opened.

In most cases if the pump is running it can be heard. However, that is not always the case. Sometimes, the pump may get very hot, so be careful when you touch it. Of course, the heating fluid will heat the pump but often it is much hotter if it is seized.

A good and very simple method of checking if a pump is rotating is to listen to it using a screwdriver, or piece of wood resting against one side of the pump (Figure 13.6).

Figure 13.6

Figure 13.7

If you are still unsure, then the best way is to check the shaft for rotation; all that is required is for the cap on the side of the pump to be unscrewed (Figure 13.7). It is a good idea to put a cloth or small container under the pump when you unscrew the cap, as it will leak a bit of water. Don't worry, though, as it will not pour out and flood the place. Once the cap is off, gently put a screwdriver in the end of the pump housing (Figure 13.8). If it is rotating you will feel it against the screwdriver. If it is not rotating, then you can move on to the next step, which is to check at the pump electrical terminals to see if there are 230 volts.

If the pump has a supply to it, then it could be one of three things and you will need to check them all.

To check to see if it is seized, put a screwdriver onto the end of the pump shaft and give it a twist. Try it a few times, as if it is seized it may take a few turns to completely free it. After a few twists of the screwdriver it should just keep going.

If it does run, turn it off and let it stop and then turn it back on again to see if it starts on its own accord. If it does not restart, repeat the process a couple of times just to ensure that the shaft is rotating freely. If the pump only keeps running when you spin it, then it is indicating that the pump start capacitor is faulty (Figure 13.9). It is a simple job just to replace the capacitor.

Figure 13.8

Figure 13.9

If the pump will not run at all, rotates freely and it has a supply to it, then it probably is a faulty pump. It is possible to check the resistance of the windings but in reality if you have carried out all of the checks described, then just replacing the pump is the best option.

Where the central heating pump is accessible it is not usually a difficult job to change one, providing of course that it has isolation valves that are working correctly. Usually, the most difficult part once it has been isolated is to release the unions to the pump (Figure 13.10), particularly if they have been sealed using boss white or similar.

When the pump is not easy to get at to work on, or you can't get the unions undone, it is often possible just to change the motor by unscrewing it from the body.

Make sure that you have the same model replacement pump as the one that is already installed. Take the motor off the new pump so that it is ready

Figure 13.10

to be fitted to the existing pump body (Figure 13.11) and that the pump is isolated from the electrical supply and that the water isolation valves are off. Put a container or towel under the pump, as it will be full of fluid and it will probably be black. Now undo the Allen screws that are holding the old pump motor in place, but do not take them completely out just in case the water isolation valves have not shut off properly. It may take a light tap with a hammer to free the motor from the body. When it begins to free just hold it to let the water out; you will soon know if the valves have held by the amount of water that flows out.

As soon as the water stops, take the old motor completely off, clean out the body if necessary and fit the new motor onto the old body, making sure that you have the gasket in place. Once you are sure that the motor is fitted correctly and all of the screws have been tightened, undo the heating isolation valves and check for leaks.

Where the pump has been fitted in a tight space and is difficult to get to it is a good idea to carry out the electrical connections to it before you fit it to the old body. Also remember that the seals in the pump are made of carbon and the pump should never be run without any water in it, as running it may damage the seals.

Figure 13.11

Motorised valves

These are valves that open and close electrically. There are two-port valves that just open to allow water to flow, and close to prevent the flow of water. These are generally referred to as zone valves (Figure 13.12).

There are also three-port diverter valves that either divert the water one way or the other (Figure 13.13) and three-port mid-position valves that divert the water one way or the other and will also allow the water to flow both ways when required. The three-port valves look identical, so be very careful to ensure you know which valve you are dealing with.

There are various things that can go wrong with a motorised valve. Some can be repaired just by changing an internal component and others will require a complete valve replacement.

Two-port valve

Most two-port valves are made to open when the motor is energised and spring back to shut when the motor is de-energised. The motor is a 230-volt synchronous motor, which is very easy to change (Figure 13.14).

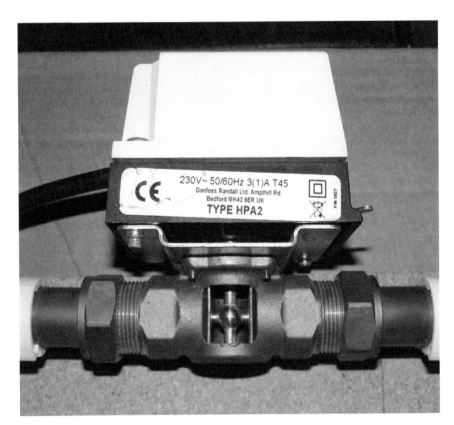

Figure 13.12 Body cut away to show how it works

Figure 13.13

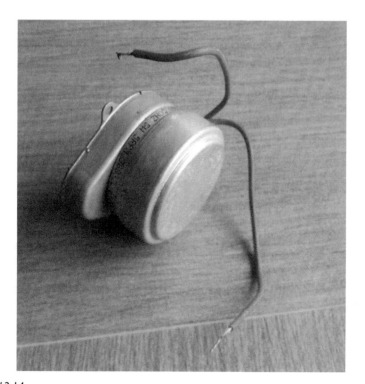

Figure 13.14

The same motor is suitable for all valves described here.

Before rushing into changing the motor, a few simple checks are a good idea. The first check is to make sure that the motor is being energised when it should be by turning on the heating or hot water and checking the voltage at the motor terminals with a voltage indicator.

If there is voltage, then clearly it is the motor that is at fault. If there is no voltage, then the fault will be elsewhere and the wiring should be traced back and the voltage measured at each point.

To change the motor it is just a matter of undoing two retaining screws and disconnecting the wires to get the motor out (Figure 13.15), and reversing the procedure to put the new motor back. The resistance of a good motor is around 2400Ω. Usually, if a motor is faulty it will just go open circuit and then when you measure the resistance it will be very high, often in megohms.

To make life easier, on modern valves it is usually possible to remove the valve head from the valve body, but be very careful as this was not possible on the older type of valves. The good news is that it is a simple task to tell the difference between the valve heads that can be removed safely and those that can't.

Figure 13.15

As you can see, on the top of the casing there is a pip (Figure 13.16). Where this pip is present, it is safe to remove the power head from the valve without the risk of flooding. If the power head does not have a pip it is always best to seek advice from the manufacturer.

- A two-port valve will have five wires
- Green and yellow = earth wire
- Blue = neutral
- Brown = motor supply
- Grey = permanent live

Figure 13.16

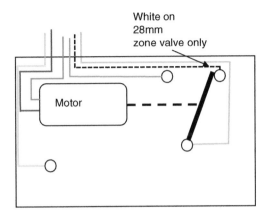

Figure 13.17

- Orange = switched line
- White = (only on 28mm bsp valve) for use as auxiliary control, as it is on the third terminal of the micro switch – it connects to orange when the valve is in the closed position (Figure 13.17)

One important thing to remember about the colours is that, although the brown, blue and green/yellow have specific purposes, the grey and orange are just the line and return conductors from a switch, which will do the job however they are connected.

The micro switch inside the valve head is normally open when the valve is at rest and only closes when the valve opens. For that reason, if it is suspected that the micro switch is faulty, and you need to carry out a dead test to check that it closes when the valve is open you have to use the manual lever (Figure 13.18).

You will see that on the lever end of the valve head that there is a notch that can be used to hold the valve open. This will *not* activate the micro switch when resting in the notch. To activate the micro switch you must push the lever as far past the notch as possible. Normally, you can hear a faint click when the micro switch turns on.

If you are trying to trace a fault on the system where the hot water will not turn off and the valve and switch appear to be operating correctly, the problem could be with the small rubber ball inside the valve body, which is used to close off the port. Occasionally, the ball gets damaged/falls apart and does not close off the port correctly.

It is also worth checking that the valve stem moves freely. It sometimes seizes and become a bit tight during the summer through lack of use, preventing the motor from opening the valve. With the body removed just turn the valve stem back and forth until it feels free.

Figure 13.18

Three-port diverter valve

The operation of the motor can be checked using the same procedure as for the two-port valve.

The valve only has three connections: a green/yellow, which is the earth; a brown, which is the supply; and a blue, which is the neutral.

When there is no supply to the brown this type of valve is spring loaded to automatically return and close port A, which is the central heating port. This, of course, will leave port B open.

The simplest way to check this type of valve is just to energise the motor and check that the spindle rotates to close the valve and then de-energise the valve to check that it closes. If the valve appears to open and close correctly but it is still letting water through, then it will be one of two things. Either the valve is not opening and closing fully, in which case the stem may be seized slightly, or the ball used to close the opening has become damaged.

Three-port mid-position valve

The same motor checks as for the two-port valve can be used to check that the motor is serviceable.

This type of valve has five connections: a green/yellow, which is the earth; a blue, which is the neutral; and a grey and a white wire, which are control wires. There is also an orange, which becomes live when heating or domestic hot water is called for; this usually supplies a feed to the pump and boiler.

With the neutral connected, if you energise both grey and white, the valve should move to port B and open port A for the central heating. When the valve opens to port B the orange wire also becomes live.

With the neutral connected, and the grey and white wires disconnected the valve should spring return to close port A and open port B for domestic hot water.

With the neutral connected and only the white energised the valve should stop in mid-position, allowing water to flow both for heating and domestic hot water. The orange should also become live.

So, in a summary:

- No power to valve, port B open for HW
- Power to white wire only, valve in mid-position
- Power to both grey and white wires, there will be an output on the orange wire
- Power to grey wire, valve held in the last position and, if the last position was CH, then there will be 100 volts of output on the orange wire.

Chapter 14

Electrical installation fault finding

As far as I am concerned, the most effective method of fault finding is to work methodically. This, combined with experience and a bit of luck, usually ends up with the correct result. Of course, not all faults are the same and some are very difficult to find, but once found they are usually quite simple to correct.

I find that it is always better to step back and think about what you are doing, and not jump in with both feet and start taking things apart. The very first step is to gain as much evidence/information as possible; ask the customer or the person who has discovered a fault to provide as much information as possible, as this will help.

I always ask if anyone else has been trying to find the fault. Has anything been disconnected? This information could have a massive effect on the way you approach a fault. For instance, if something has been working, but will not work now and has not been touched other than to switch it on and off, that indicates it is a genuine fault and that something in the circuit has failed. If, however, someone has been pulling it around and disconnecting things that they know nothing about, you will have a different job on your hands, as you cannot be sure that it is connected correctly.

Ring final circuit

Let's say we have a ring, which, when we measure end-to-end values we notice has no continuity between the ends of the line conductor but all of the other conductors measure are as expected.

Of course, the break could be at any point in the ring and is probably just a cable snapped at one of the terminations. As with most fault finding, we need a bit of luck mixed with our knowledge. The most important thing to remember is to work methodically.

It could be that we have not installed the ring and have no idea as to the route of the cables. We now have to just take a calculated guess as to how the ring is wired.

Take a socket off somewhere around the middle of the ring.

- At the board, link L and E of one end of the ring.
- At the socket, test between L and E of one of the ring ends using a low resistance ohm-metre.
- If you have continuity, go to the board and split the joined ends, but mark which ones they were.
- Now test L–E again at socket and if the circuit is open resistance you will know that the break is not on the end of the ring which you have tested. Again, mark the end that is OK.
- Now join the L–E of the other end of ring at board.
- Test L–N at the socket of the unmarked end and the circuit should be open circuit.
- If it is open circuit, join the L–E of the end just tested.

Now comes the tricky bit, as you have to guess which part of the ring has the break. You need to take a chance on which way the ring has been run, and take a socket off somewhere between the board and the socket.

- Disconnect conductors from the socket.
- Test between L and E of both cables.
- If both ends are open circuit there is a good chance that you are working on the wrong side of the ring. The best way to find out is to join L–E of the ends that you marked at the socket and board, test again at the socket and you should have a closed circuit. If that is the case, then just put the socket back on and choose another somewhere else on the ring.
- If, however, when you take the socket off and measure the ends one of them is closed circuit and the other is open circuit, you will be looking in the right place.
- Now you need to guess which way the cable that is open circuit runs; it may go to the board or it may go to the first socket that you removed.
- Take another socket off between the socket you have just tested, and either the first socket or the board. This is your choice and also a bit of luck is needed.
- You need to keep doing this until the fault has been found; it is just a process that keeps halving the faulty section.

Of course, once the fault has been found, then all of the connections need to be remade and a complete-ring final circuit test will need to be carried out.

Interpretation of ring final test results

Let's assume that we have a ring circuit wired in 2.5mm^2 thermoplastic 70°C twin and earth cable. The cpc is going to be 1.5mm^2.

The test between the ends of the line conductor gives a reading of $0.6\,\Omega$. From this value we will know to expect a value of $0.6\,\Omega$ when we measure the N conductor end to end and a value of 1.67 times $0.6\,\Omega$ ($0.6 \times 1.67 = 1\,\Omega$) when we measure the cpc end to end. If the values differ from these by very much, then immediately we should recognise that there is probably a loose connection (high resistance joint) in one of the sockets. Before we proceed with the rest of the test we should correct the problem.

Once corrected, we will have values of R_1 $0.6\,\Omega$, R_n $0.6\,\Omega$ and R_2 $1\,\Omega$.

When we cross connect the line and N conductors we know from these values that we will expect a resistance value between L and N at each socket of $\frac{0.6+0.6}{4} = 0.3\,\Omega$. In reality, though, we will know that the expected resistance value will be half of the line conductor resistance.

When we cross connect the line and cpcs we will expect a value of $\frac{0.6+1}{4} = 0.4\,\Omega$. Always remember, though, that as the cpc is a smaller CSA than the line conductor, the value will be slightly lower on the sockets nearer to the cross connection but will gradually increase to $0.4\,\Omega$ as you measure the sockets nearer to the centre of the ring.

In Table 14.1 are some measured values that, if interpreted correctly, will give us a good idea of how a problem could be identified.

Now let's look at how these values have been interpreted to find the possible faults/inconsistencies. Before we do, though, I should say that if you ever install a ring that has all of the faults listed above, you should seriously consider becoming a plumber!

Table 14.1

L–L = 0.6Ω	N–N = 0.6Ω	cpc–cpc = 1Ω	
	L–N cross connected 0.3Ω	L–cpc cross connected 0.4Ω	
Socket	L–N	L–E	
1	0.3Ω	0.4Ω	correct
2	No reading	0.4Ω	Reverse L–E
3	0.6Ω	0.4Ω	High resistance N connection
4	0.6Ω	0.8Ω	Possible spur or high resistance L connection
5	0.3Ω	0.7Ω	High resistance E connection
6	0.3Ω	No reading	Reverse L–N
7	No reading	No reading	Reverse N–E

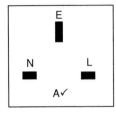

Figure 14.1

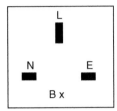

Figure 14.2

Socket 1

We can see that the values are as expected (Figure 14.1).

Socket 2

As we look at a socket face (A) we expect the connections to be as shown as in Figure 14.2.

If the connections were as shown in (B) the reading when measured between L and E would be as expected but would be open circuit between L and N, proving that the socket has been connected incorrectly or the N conductors have dropped out of the terminal.

Socket 3

The value shows us that it must be a loose N connection as L–E are as expected (see Figure 14.3).

Socket 4

As the values are both high it will show either a spur or a loose connection on the conductor which is common to both tests. This, of course, is the line conductor (Figure 14.4).

Socket 5

If it is OK between L and N but high between L and E it must be a loose earth connection (Figure 14.5).

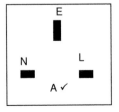

Figure 14.3

Figure 14.4

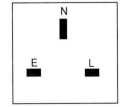

Figure 14.5

Socket 6

No reading between L and E but a correct reading between L and N will show that L–N are reversed as shown in socket (B).

Socket 7

If there is no reading at either measurement it must show that the NE are reversed as shown.

All of these mistakes are very easy to make, and it is a good reason to carry out all tests correctly as human error is very difficult to eliminate.

RCD tripping

If the circuit is protected by an RCBO, then, of course, the first step is to disconnect as much of the equipment connected to the circuit as possible. If it is a power circuit, then unplug everything that you can find, switch off fused connection units, then try and reset the RCBO. If it will not reset, isolate the circuit and disconnect the line conductors but not the cpc. The cpc should be left connected to the earth bar, as the fault may be between the circuit live conductors and another exposed or extraneous conductive part, not necessarily between the live conductors and the cpc.

Now carry out an insulation resistance test between the live conductors joined together and the earth bar. I usually carry out the test at 250V first just in case there is surge protection. Now split the circuit around the centre. This will indicate which side of the circuit has the fault. Split the faulty side of circuit again until you have found where the fault is. Of course, it is not always that easy, as the fault may be under the floor (crushed cable, rodent damage) but at least the test will narrow down the area for you.

If the fault is on a lighting circuit, switch off all of the lights and then try and reset the RCBO. If it resets, then just switch on the lights one at a time until you find the one that trips the RCBO.

If, however, the RCBO will not reset, then have a good look around for any switches that look like they may have been disturbed, or light fittings that have been recently fitted. If there is nothing obvious, then isolate the circuit and disconnect the live conductors. Do not disconnect the cpc.

Where installations are protected by an RCD main switch, the obvious process is to switch off all of the circuit breakers or remove all of the fuses. Switch on the RCD and replace the protective devices one at a time until the RCD trips, remove the device that trips the circuit and replace all of the others just to be sure that the fault is only on one circuit. This will then identify the circuit and then you can begin the fault finding.

If the RCD will not reset when all of the protective devices have been removed, it will indicate that the fault is a neutral to earth fault. The procedure now is to remove the neutrals one at a time and carry out an insulation resistance test between the disconnected N and earth until the faulty circuit is found. Once found, replace all of the other disconnected neutrals and protective devices and switch on the RCD. If it stays on, then you have found the faulty circuit.

Something to be aware of when fault finding and RCDs are involved is a situation that often arises when RCBOs are used along with an RCD main switch. Occasionally, the main RCD will trip, as well as the RCBO, even if the main switch is time delayed or higher rated.

This is because most RCBOs are single pole and although they identify an N to E fault and disconnect, the N remains connected and in turn causes the RCD to trip.

When upgrading consumer units to comply with BS 7671 and a split board is used with two or three RCDs, it is important to ensure that the neutrals and line conductors are connected to the same section of the board. If a mistake is made you will soon discover that when something is connected to the circuit the RCD will trip – this, of course, will look like a faulty circuit.

Insulation resistance

When insulation resistance testing, always take care as the test is usually carried out at 500V DC. The test leads and probes/clips should comply with GS38 and it is really important that the tester is checked for correct operation by connecting the leads together and pushing test button. This should provide a reading of 0.00. Now part the leads and repeat the test and a reading of the maximum resistance should be seen such as >300MΩ, but this will depend on the tester being used.

Remember, that this is a dead test and the circuit or complete installation must be isolated before testing can begin.

Let's assume that we are going to test a complete installation for a condition report and that isolation has be achieved by switching off at the main switch.

Now we must look around the building for any fixed appliances that may be switched on – any found should be switched off.

Unplug anything that is plugged into a socket outlet. Isolate any controls for the heating/hot water system; this is usually just a matter of switching off or unplugging the heating control supply.

Check that any neon indicator switches are turned off, as well as any equipment that may be damaged by the test voltage; this could include motion sensors and dimmer switches. These should be either bypassed or linked across.

Remove any lamps where possible. If discharge lamps are used, such as fluorescent fittings which have control gear fitted, or even any lamps that are difficult to access, just turn off the switch that would operate them.

What we have done is to try and remove any equipment that may result in giving us a low reading due to our test current passing through it.

With the main switch off and all of the circuit breakers on or protective devices in place, we can begin the test.

The first test that I would carry out is between live conductors and I would always start at 250V. This is because any equipment that we have missed but that could be damaged at 500V will show up as a bad reading but will not be damaged.

If the test results in a low reading, then I would look around again just to check for anything that has been missed; a simple method to narrow the search would be to turn off all of the circuit breakers or remove protective devices. Carry out the test again but switch on the circuit breakers/replace devices one at a time; this should identify the circuit with the problem.

Identifying a lost switch line on a three-plate lighting circuit

I am sure that I am not the only person to get numerous calls throughout the year after someone has taken down a light fitting, and finds that when they reconnect it there is a bang! The story is usually that they have connected all of the blues/blacks together, and all of the brown/reds together, and when the light is switched on all of the other lights go out.

I find the simplest method is this.

Isolate the circuit (*it probably already is, but by accident*). Don't ever take anyone else's word that a circuit is dead. Always carry out the procedure yourself and always lock off correctly. Better safe than sorry!

Once isolated, disconnect all of the blues/blacks, turn the switch on and connect a buzzer or low resistance ohm-metre to the browns/reds and one blue/black. Test each blue/black to the brown/reds until you get a reading. Once you have a reading you must now turn off the switch. If the reading disappears you have found the switch return. If it does not disappear you will most likely have found a live pair coming from another light, in which case move on to another blue/black.

Example

I was once asked to repair some swimming pool lights that had stopped working. I followed the normal procedure of finding out as much information about the lights as possible, and when they stopped working. Gaining as much evidence as possible is vital.

The information was:

- The lights were quite new and were LED colour-changing lights controlled electronically by a remote control.
- There were three underwater lights, which were supplied by a transformer with a 12V AC output.

- The transformer was fed from by dedicated circuit, which had its own control switch on the pool switch panel.
- The client said that the lights had previously worked but only for a very short time.
- Each light was fed by its own cable, which was run into a trunking.
- The cables were joined together in the trunking using a crimped connector. A single cable was taken from the crimp to the supply transformer.
- It was not possible to gain access to the lights without draining the pool.

To find the fault: I checked the supply to transformer. This was OK, as it measured 230V. Next, I checked the secondary side of the transformer, remembering it is AC. It had an output of 12V.

My first thought was that it was the lamps that had gone faulty, but it was pretty unlikely that all of them would be faulty and the only way to find out would be to drain the pool.

Then, I checked the voltage at the crimp, which again measured 12V. Before draining the pool I thought that I would put a load test on the transformer. I connected a 12V car lamp to the transformer and it worked. This at least proved the transformer.

I was still not convinced that the lamps were faulty and decided to split one of the cables between the crimps and lamp, tested between live conductors and there was no voltage. This seemed strange, as there was voltage up to the crimps, so I decided to split the crimps carefully. I found that on the outgoing side of the neutral the crimp was squashed onto the insulation; the copper wire was not touching the metal of the crimp and therefore was not completing the circuit.

The lamps had worked for a short while because the wires were twisted together with one strand longer than the others. When it was first crimped the single strand of the outgoing cables was touching the incoming cable. This single strand could carry the current for a short while, but when the lamps were left on for a long period it acted like a fuse on overload and melted, breaking the circuit.

This was a pretty obscure and confusing fault, but, by working methodically and not jumping around all over the place, the fault was found reasonably quickly.

Immersion heater not working

The faults on immersion heaters are usually quite easy to identify, although there are many symptoms and various ways of identifying the faults.

Where the circuit is protected by an RCD, the breakdown of an immersion heater element will usually result in the RCD tripping. The most common cause of element breakdown is the corrosion of the protective tube around the heating element; when the tube ruptures water can penetrate and will soak into the magnesium oxide surrounding the element. This soaked magnesium oxide will provide a low resistance between the element and the outer protective case, which of course will be earthed.

The easiest way to prove this fault is to just to turn off the local isolator to the immersion heater and then turn on the RCD. If it stays on, then it will show that the immersion heater is the problem and will need to be replaced.

Always be aware, though, that the isolator needs to be double pole. If it is not, then the RCD may still trip as it will detect the leakage between earth and neutral.

Where an RCD is not part of the circuit a damaged element will not usually operate a circuit breaker, and the only symptom may be that the immersion heater is not working. This type of fault will require a methodical approach.

Always check the obvious. Is the circuit switched on? If it is connected by a switched fuse connection unit (fused spur), is the fuse in the connection unit the correct size? Is it in good condition? If yes, then switch off the local isolation to the heater and remove the cover of the immersion heater.

Visually inspect the connections in the immersion heater just to ensure that the connections have not overheated or come loose.

If it has an integral overheat trip, press the trip button (Figure 14.6). If it clicks it will mean that it has reset.

Figure 14.6

Figure 14.7

Now turn on the local isolation and, using a voltage indicator to GS38, test between the ends of the element within the immersion heater (Figure 14.7).

If the voltage indicator shows that the full voltage is present, then it will indicate that the thermostat is closed and that the element has a supply to it.

If there is no supply to the heater element, then test between the incoming supply to the thermostat and neutral or earth (Figure 14.8). If there is a supply voltage, then at least you will know that the circuit is working up to the thermostat. This will indicate that the heater thermostat may be faulty or perhaps has been turned down. Check the setting just in case – it should be at a maximum of 60°C. If the water is hot because it has been heated by a boiler it may be that the temperature stat has turned off.

It worth just turning it to full temperature and then testing at the heater element ends again. If they are still showing as dead, then replace the thermostat.

If the ends are live, then clearly there is a problem with the element.

At this point we have a couple of options:

Switch the heater off and place a clamp meter around the line conductor to the heater or the incoming line at the meter (Figure 14.9). Now switch the heater on; if the current showing on the clamp meter increases by the current rating of the element it will show the heater is working correctly. A 3kW element will increase the current by 13A. You can also carry out this test at the meter tails if the immersion is difficult to get to. Just test it with the immersion off and then turn the immersion on to see if the current increases by the current rating of the immersion.

Figure 14.8

Figure 14.9

Figure 14.10

If the current does not increase and you have a supply to the element it will show that the element has gone open circuit.

As a final check it is worth isolating the element and carrying out an insulation resistance test between one of the element ends to earth (Figure 14.10). A low reading will show that the element case has eroded away and allowed water to get into the magnesium oxide insulation surrounding the heater element.

Changing/replacing an immersion heater

Once you are sure that the immersion heater needs replacing, then the removal of the old heater is usually the most difficult part of the exercise, as the cylinder will need to be drained.

There are many scenarios and it would be impossible to describe them all. The next chapter will deal with top and bottom elements in a standard vented cylinder. I will also describe the procedure for replacing an element in an unvented cylinder.

Before we start, be warned that there is always a danger of damaging the cylinder when removing the old element, particularly if the cylinder is an old one.

Chapter 15

Changing an immersion heater

Before you start read this chapter, and if you are concerned, employ the services of a plumber.

- Isolate and disconnect the electrical supply to the immersion heater.
- Locate the cold feed into the cylinder (this is usually the 22mm pipe that enters the cylinder at its lowest point). See Figure 15.1.
- Once the inlet has been located trace the pipe away from the cylinder until you reach a valve; this is normally a gate valve (Figure 15.2). Once you have located it try and turn it off; very often, the gate valve will be seized up so be careful. You can use a pair of grips for leverage but do not apply too much pressure; the valve stem is made of brass and will shear easily.

Figure 15.1

Figure 15.2

- Unfortunately, it is not uncommon for the valve to shut but then seize shut, which presents an even worse problem. If there is any doubt that the valve is serviceable, then don't force it.

If you are concerned about the valve it will probably be better to drain the cold water storage tank, which is usually located in the loft. This will be where the water is stored to supply the hot water cylinder and cold taps in the bathroom.

To drain the cold water tank there are several methods that can be used.

- **Method 1.** Go into the loft and turn off the stop cock in the cold supply to the cold water storage tank. If there is no valve, tie up the ball valve, which is in the cold water storage tank. Isolating the tank supply this way will allow the cold water to remain on to all taps supplied by the main water supply.
- **Method 2.** Turn off the main stop cock in the pipe supplying the building (Figure 15.3), or turn off the supply by using the stop tap, which is in the box on the pavement (Figure 15.4). This may need a simple square key or a socket with a long extension (Figure 15.5).
- Once the cold supply to the cylinder has been isolated, open the hot taps in the kitchen and the bathroom. If it has been possible to isolate the hot water cylinder locally there will not be much water flowing out of the taps and it will stop quite quickly.

Figure 15.3

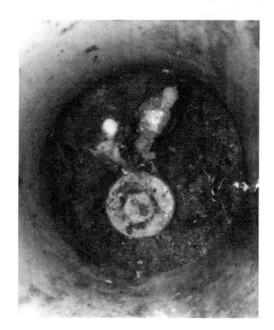

Figure 15.4

Figure 15.5

Figure 15.6

If you have isolated the cold water storage tank the water will flow for a while as the storage tank will have to empty. Opening the cold taps in the bath will speed things up a bit as this is usually supplied from the cold storage tank as well.

Once the water has stopped flowing from the tap it will mean that all of the pipes are empty, but *not* the cylinder – this is the next step.

- At or near the bottom of the hot water cylinder there should be a drain off valve. It is often fitted to the cold supply pipe to the cylinder as in Figure 15.6.
- Fit a hose over the end of the drain off valve. If the immersion heater is on the top of the cylinder you will not need to drain much water out and you can just put the end of the hose into a bucket. Make sure you put a towel or a tray below the drain off as it will leak a bit.
- Where the immersion heater is fitted to the bottom of the cylinder the water from the cylinder must be drained to below the immersion heater. This will require the end of the hose to be lower than the immersion heater.

Often, particularly on older cylinders, the washer in the drain valve gets stuck to the seat of the valve and will not allow any water to flow out. This can be a pain to deal with.

You can completely unscrew the valve (Figure 15.7). Dig the washer out and block up the hole using your thumb (Figure 15.8). If you use this method make sure you have someone with you just in case the hose moves

Figure 15.7

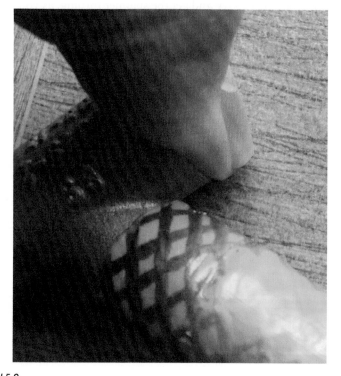

Figure 15.8

or you need help with anything. Remember that once you have taken out the washer the cylinder will need to drain down completely before you can take your thumb away. If the water is still hot you will need to use a rag or towel to block the hole.

On occasion, it may be impossible to drain the cylinder from the bottom. In these cases, I undo the hot water outlet pipe at the top of the cylinder and siphon the water from the cylinder. This is a simple procedure providing you think about what you are doing.

- Make sure you have a long enough hose to be able to siphon the water to a point below the cylinder, preferably outside into a drain.
- Fill the hose with water and then block the lowest end of the hose so that the hose remains full up.
- Push the hose into the top of the tank and get the end as close to the bottom of the tank as possible.
- Unblock the other end of the hose and let the water siphon out.
- Once the water is below the immersion heater you can stop draining down.
- It is always better to use an immersion heater spanner to undo the immersion heater (Figure 15.9).
- It is often difficult to get enough leverage on an immersion heater spanner; where there is enough space a steel tube placed over the end will usually help. A pair of 24-inch Stilsons will also do the job when opened up fully.
- Do not use a hammer to try and get the spanner turning, as you will end up damaging the cylinder.

Figure 15.9

Unfortunately, on old cylinders where the copper has corroded inside, the thread which the immersion heater screws into will often split where it joins the cylinder. This can't be avoided and when it happens the cylinder will require replacement.

Once the heater is out, the replacement is quite simple to fit. When I fit immersion heaters I try not to use a sealant that sets, as it makes the removal of worn-out heaters very difficult. The fibre washer that is supplied with the heater is normally good enough to get a good seal. If in doubt, try and use a sealant that remains soft.

- Ensure the surface of the boss on the cylinder is clean and smooth.
- Fit the fibre washer over the thread of the heater and then screw the heater into the boss.
- If you have to bend the element a bit don't worry, as it will not harm it.
- Tighten the immersion heater using the correct spanner; be careful not to damage the cylinder.
- Once you are sure the immersion is fitted correctly, turn off all of the taps except the bath tap.
- If the drain valve on the cylinder was used, check the washer and replace it if necessary.
- Where the cold storage tank has had to be drained because the gate valve to the cylinder supply did not work, it would be sensible to replace the valve while the system is drained down.
- Once you are happy that the drain off is shut and all disconnected pipes are connected, it is time to open up the valves to fill the cylinder or refill the cold water storage tank.
- Where the storage tank has been drained the cylinder will fill as the tank fills.
- While it is filling you can reconnect the electric supply to it. Check that the flexible cord does not need replacing. If it does, be sure to use heat-resistant cord. Where possible use crimped lugs or ferrules on the end of the cord.
- Set thermostat to 60°C.
- Once the cylinder has filled, check for leaks and switch on heater. It will take a while to heat up but just to check it is working you can use the clamp meter on the meter tails to check the increase in current.

Unvented cylinder

Where there is an immersion heater fitted into an unvented cylinder the tests carried out to check that it is functioning are much the same as for a vented cylinder.

Always check that there is a supply at the element ends and then just work back from there, but be sure that you follow all of the safety precautions regarding safe isolation.

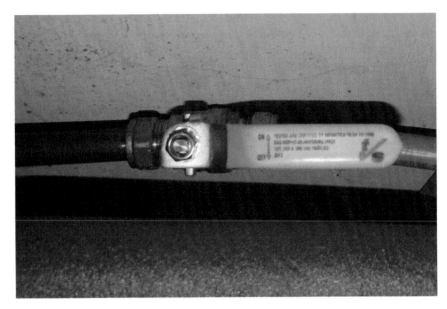

Figure 15.10

Changing an immersion heater on an unvented cylinder is usually a lot easier than changing one on a vented cylinder. This is because the isolation valve is normally a lever type that is situated close to the cylinder (Figure 15.10).

The cold fill pipe is normally the one that enters the cylinder at the lowest point. Quite often, this is where you will find the drain off valve as well (Figure 15.11).

Before you start, make sure you switch off and isolate any heating system that may be used to heat up the cylinder. This is because a heating system will have some electrical safety features to prevent the cylinder overheating. The wiring may pass through parts of the cylinder and will be live even if the immersion heater is isolated.

Now isolate the immersion heater. Once that is done turn off the water isolation valve and open a hot tap. This will be under pressure for a short while as the expansion vessel will be pressurised.

After a very short time the water will stop running. Now connect a hose to the cylinder drain off point, put the other end of the hose outside or in a sink/WC that is lower than the drain off point and undo the drain off valve. Let the water run until it stops.

This may take a while, particularly on cylinders that have been installed for a while and those that are in hard water areas. The limescale can sometimes build up to a level above the drain off and partially block it. When it is completely blocked, the siphon method previously explained for a vented system will work well.

Figure 15.11

Once it is drained, remove the immersion heater using an immersion heater spanner. If there is a lot of limescale in the cylinder now is a good time to remove it using something to spoon it out through the immersion heater hole.

Clean the surface of the immersion heater boss and install the new immersion heater and tighten using the immersion heater spanner. Try not to use any sealant that will set, as it will make replacement more difficult next time.

Now turn the water supply back on leaving the hot tap open. While the cylinder is filling reconnect the immersion heater and set it to 60°C. Once the cylinder is full, water will come out of the hot tap. When this happens, turn the hot tap off.

Check for any leaks. If there are any, they will show now as the cylinder will be under pressure. When you are satisfied that the cylinder is not leaking, check that the immersion heater is working by using a clamp meter, or just wait a short while to see if the outlet at the top of the tank is getting warm. Once you are satisfied that it is working, re-energise any heating system that has been switched off and check that it is operating correctly.

Chapter 16

RCD tripping faults

It is rare these days for RCDs to cause nuisance tripping. Usually, when an RCD trips it is because something is wrong. It is unlikely to be the RCD that is causing the problem.

One important thing to remember is that an RCD will only trip on a fault between live conductors and earth. It will not trip when there is a fault between live conductors or when there is an overload.

Where the RCD is part of a split board and is a new connection, it is always worth checking that the neutrals of the circuits are all connected to the correct neutral terminal. In other words, check that all of the circuits that are protected by the RCD that keeps tripping are connected to the neutral terminal corresponding to the side of the board that the RCD is protecting.

If a circuit neutral is connected wrongly, and a load is connected to the circuit, the RCD will detect more current in the neutral than the line conductor, as the neutral conductor and line conductor pass through it and trip.

Where everything seems to be connected correctly, it is worth carrying out what is called a ramp test before starting to take things apart. A ramp test will tell you how much current is passing through the test instrument before the RCD trips. Clearly, if a 30mA RCD trips at a current of, say, 20mA it could be prone to nuisance tripping.

The problem with this test is that if it is carried out on a circuit while all of the other circuits on the board are still connected, any small earth fault currents on other circuits will have an effect on the RCD. Although the test instrument is showing 20mA it is only measuring the current that is passing through the test instrument. It will not be able to measure any leakage current that is in the rest of the system.

Before you change the RCD, isolate it from all of the other circuits and test it on its own. If it still trips at 20mA, then change it. As a precaution I would always test the rest of the circuits on the board between live conductors and earth to check if there are any low readings, just to satisfy my own mind that I have cured the fault.

Where the RCD is part of an older circuit that has been working for some time but has just started tripping, it is unlikely to be the RCD causing the problem. Once again, the very first thing I would do is gain as much information about the fault as possible. Ask whoever is using the installation questions as these will usually give you some clues as to what the problem is.

Always remember that many people have no idea at all of the way any electrical systems work. Don't be embarrassed about asking the obvious, even if the person seems to know what they are talking about.

- How often does it trip?
- Did it trip just once and won't reset?
- Did it reset a few times before it began not resetting at all?
- Does it trip when you plug something in or switch a particular part of the circuit on?
- Has any work been done around the house: shelves put up, floorboards screwed down?
- Have any light fittings or switches been changed?
- Have you plugged in anything new recently?
- Has there been a water leak anywhere recently?

Once you have asked the questions, then it's time to have a bit of a poke around.

The first objective is to try and detect the circuit that is causing the problem. Of course, if the board has RCBOs then the circuit causing the problem will be obvious. Always remember, though, that most RCBOs are single pole, and when you switch them off or they trip, the circuit may be off but if the fault is between neutral and earth the fault will still be there.

Where the circuit is part of a system that is protected by a single RCD, or even part of a split board, then the problem circuit will not be obvious.

Switch off all of the circuits that are protected by the problem RCD. Now switch the RCD on. If it stays on, then turn on the circuit breaker nearest the RCD. If it stays on, leave it on and switch on the next one. Keep doing this until the RCD trips. With a bit of luck the circuit breaker that you switched on just before the RCD tripped will be for the problem circuit.

However, don't get overexcited as the problem may be a build-up of small leakages over several circuits. This is simple to check – just turn all of the circuit breakers off and switch on the RCD. Now switch on the circuit breaker that caused the RCD to trip; if the RCD trips again you have found your problem circuit.

On the rare occasion when the RCD does not trip it may mean that more than one circuit is involved. Starting with the circuit breaker nearest to the one that you have switched on, switch the circuit breakers on one at a time until the RCD trips.

Now turn off all the circuit breakers other than the first and last ones that you switched on and try to reset the RCD. If it resets, then switch on another breaker. Keep repeating this process until you have found the combination of circuits that trips the RCD and those circuits which don't.

Disconnect the live conductors of the problem circuits one circuit at a time (not the cpcs). Now carry out and insulation test between the live conductors joined together and the earth bar. This will show you which are the problematic circuits.

Where you find it is only one circuit causing the RCD to trip, things are slightly easier.

The RCD is acting as a main switch and it trips but will not reset.

- Switch off all circuit breakers or remove all fuses.
- Reset the RCD.
- If it resets, switch on the circuit breakers or replace fuses one at a time. When the RCD trips you have found the problem circuit.
- If the RCD does not reset, this could indicate an N to E fault.
- Remove the neutrals one at a time; each time one is removed, try and reset the RCD. If the RCD does not reset, do not reconnect the N but disconnect another one.
- Once the RCD resets, mark the last N disconnected for identification purposes.
- Reconnect the other neutrals one at a time, ensuring that each time you reconnect a neutral the RCD is switched off for isolation purposes.
- The fault will be on the circuit where the identified N is used.
- If with all of the neutrals disconnected and all of the circuit breakers turned off the RCD will still not set, it will indicate that the RCD is faulty. However, it is always a good idea to carry out a test on the RCD with nothing connected to it just to be sure.

Always be aware that it may not be one circuit that is causing the problem. If, for instance, the RCD has a trip rating of 30mA it could be that a number of circuits have a very small leakage. When all of the circuits are operational the accumulative value of the leakages may well be more than the trip rating of the RCD.

This is often the case where there are a lot of heating elements being used in an installation. It is not unusual for these elements to have very low insulation resistance values, particularly if they have not been used for a while. The metal tubes in which the element is contained can have very small hairline cracks that allow moisture to be absorbed by the insulating material, which is usually magnesium oxide. This is very hydroscopic and readily absorbs moisture.

Heating and cooking equipment with a rating $\geq$ 3kW are permitted to have an insulation resistance value of 0.3mΩ.

Where a number of heaters are being used, it is often beneficial to run them for a while to dry them out and then test them again.

RCBO trips and will not reset

- Check that neutrals are connected to the correct side of board if it is a split board.
- If the RCD is protecting a power circuit isolate all loads. If it is protecting a lighting circuit, switch off all lights.
- Try and reset the RCD. If it resets, re-energise loads one at a time. When the RCD trips you will have found your problem. If the RCD is on a lighting circuit, switch the lights on one at a time until the RCD trips. Again, you will have found the problem.
- If the RCD will not set, disconnect the line and neutral of the circuit (do not disconnect the earth) and carry out an insulation resistance test between the joined live conductors and earth.

Power circuit

Check all socket outlets and unplug anything that is plugged in, switch off/ isolate any fixed equipment such as electric fires, and make sure showers, cookers and the like are isolated from the circuit.

Now switch the circuit back on and hope that the RCD remains on. If it does, then you are well on the way to solving the problem. If it is a socket outlet circuit, leave the circuit on and start plugging things back in. When the RCD trips you have found your problem.

If it is a radial circuit supplying a piece of equipment such as a cooker or shower, then isolate the equipment and test it with an insulation tester between live conductors and earth.

On circuits that trip when everything is disconnected, then clearly there is a problem with the circuit. Now visually inspect as much of the circuit and accessories as you can see to check for anything that looks like it could be a problem. Look for damaged cables/accessories.

Disconnect from the supply end the live conductors only, and carry out an insulation test between live conductors and the main earthing bar (Figure 16.1). Of course, you are expecting a low reading below, say, 0.008MΩ because this is the value of resistance that will cause a 30mA RCD to trip.

Example:

$$\frac{230\text{V}}{0.03\text{A}} = 7666\Omega$$

which, of course, is as near to 0.008MΩ as you can get.

But, of course, anything below 1MΩ will require attention.

Remove the accessory at the end of the circuit and check for damage to the conductor insulation/pinched cables, disconnect the accessory and separate the conductor ends. Now carry out the insulation test again.

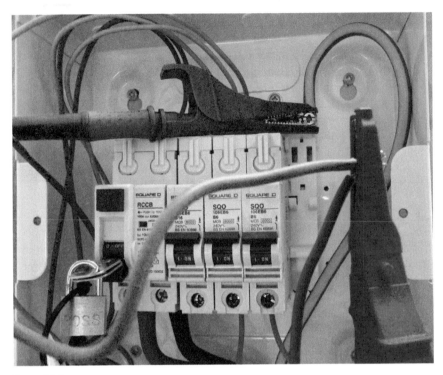

Figure 16.1

If the resistance values have increased to an acceptable level, then you have found your problem.

If you have both ends of the cable disconnected, with the conductor ends separated, and are still getting a low reading, then, of course, it is showing you that the cable is damaged. If it is a radial circuit supplying only one item of equipment, the only option is to try and trace the circuit cable and perhaps cut into it around the halfway point and test each way to see if you can narrow down the point of the fault.

If it is a radial or ring final circuit, things become a little easier. Split the circuit somewhere around the centre of it. Of course, you can only guess at where the centre is but anywhere will do to start with.

Make sure the ends of the circuit are separated at the supply point and at the point at which you are beginning your search.

Now test between live conductors and the cpc on all of the cables where you have split the circuit. One of them should provide you with a good reading and the other with a low reading. At the supply point test between live conductors and earth. If the reading is still low, then the fault is between the separated point and the supply point. If the reading is clear, then the fault is between the separated point and the end of the circuit.

Now split the circuit again between the two ends of the circuit that has the fault and repeat the process as before. Keep doing this until you have pinpointed the section of cable that has the problem. Once you have found the section of cable it is just a matter of trying to find the problem or replacing the cable.

Lighting circuit

Where possible switch off all of the lights. When you have done that, try and reset the RCD. If it resets just turn the lights on one at a time until the RCD trips. When it trips you will have found the part of the circuit that the fault is on – now you just have to locate the problem.

If the RCD does not reset with all of the lights off, carry out a visual check on everything, paying particular attention to anything that is outside and could have moisture in. If a visual check does not produce a solution, then turn all of the light switches to the off position, disconnect the live conductors at the supply end of the circuit and carry out an insulation test between live conductors and earth. It is worth switching any two-way or intermediate switches just to make sure that they are in the off position. If the reading is satisfactory, say above 1MΩ, then turn each light on one at a time while carrying out the test. When you get a low reading you will have found the problem area.

When the resistance value is low with all of the light switches in the off position, then you must follow the same procedure as for the power circuit. Lighting is often easier to work with than power as splitting the circuit is a simple process.

Split the circuit somewhere near the middle, test both ways and continue doing that until you have found the part of the cable that is giving the problem.

Chapter 17

Electric motors

As a general rule, electric motors are very reliable. Occasionally, though, they do develop faults, particularly where the motor is working close to its power limit. Before we look at the fault-finding side of motors, it is worth mentioning that it is a good idea to try and identify the cause of the problem. Although motors do sometimes just wear out, in most instances there will be a reason.

Common problems are:

- Has the motor got suitable ventilation? Check to see if the end cover is blocked. Is the fan blade able to rotate? Sometimes, they come loose. Is the motor in an enclosure of any kind? If so, has it got suitable ventilation?
- Check the voltage – a drop in voltage will result in an increase in current. As a simple example: a single-phase motor with an output of 1kW and a power factor of 0.85 will draw a current from the supply of $\frac{1000}{230 \times 0.85} = 5.11\text{A}$ from the supply.

If the voltage drops to say 200V, then the motor will draw $\frac{1000}{(200 \times 0.85)} = 5.88\text{A}$. This, of course, will make the motor run hotter and reduce its life expectancy. The same will apply to three-phase motors and, of course, the larger the motor, the greater the current.

- If the motor is three-phase, then check that all three phases are present at the motor terminals.

If there is voltage, check each phase to earth to ensure that the voltage is the same on each phase. An imbalance will cause the motor to run hot. If there is only voltage on two phases or an imbalance, check at the motor starter. Measure the voltage at the supply side of the starter. Obviously, if the voltage is correct, then the problem is with the starter. Sometimes, particularly on a motor circuit that is starting and stopping frequently, the contacts of the starter burn and cause a high resistance, which in turn affects the voltage.

- Check that the equipment which the motor is providing power to is running freely, because a piece of seized or tight equipment will cause the motor to overload.

A good sign of the equipment seizing up is if the motor starter has begun tripping. Usually, this is because the motor is overloading due to a problem with the equipment, not because there is a fault with the motor. Never just increase the overload settings on the starter, as this will just add to the problem. It may be a quick fix but the motor will get hotter and eventually burn out.

- If it is possible to disconnect the motor from whatever it is that it is driving, check that the spindle rotates freely.

If the spindle does not rotate freely, then check at the fan end of the motor to see if the fan blade has jammed or is jamming.

If the problem is not there, then it could be that the motor bearings are seizing; this usually results in a rumbling sound when the motor is spun. The fan jamming or the bearings seizing are both very simple to detect.

While the motor is disconnected I would carry out the tests as described next, anyway.

Just place a long screwdriver onto the bearing housing the motor. Place the handle of the screwdriver against your ear and rotate the spindle. You will hear if the bearings are wearing; good bearings make virtually no sound at all.

If you are concerned that a three-phase motor is just not sounding right, perhaps it sounds like it is working too hard, it is a good idea to use a clamp meter to check that the current drawn by each phase is the same. The best place to check this is at the motor starter. Set the clamp meter to a suitable current rating. Start the motor and place the meter around each phase in turn (Figure 17.1). They should all be carrying the same current or very close to the same. If the readings are different, then there is something wrong; it could be that the motor windings are burning out, or possibly a problem with the starter contacts. It will not be that the motor is overloaded, as even under those circumstances the motor will have the same current on all three phases.

Once all of those things have been checked, we can look at the motor.

Once it has been isolated and disconnected, a good indication (and often the best) is to just put your nose near the terminal box and give it a sniff. An acrid burnt smell will tell you the motor windings have overheated and melted the varnish that the windings are coated in.

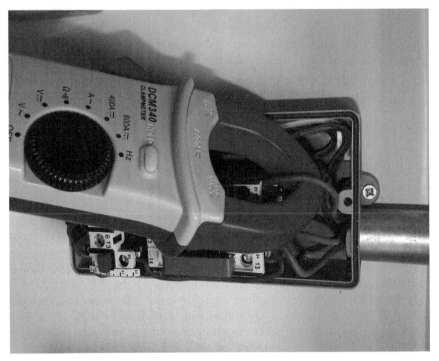

Figure 17.1

A three-phase motor will normally have three windings, and each winding will have two ends (Figure 17.2). Motor windings are usually marked V1 V2 – U1 U2 – W1 W2 (three windings marked at each end with the same letter). Three-phase motor windings can be connected in star or delta, sometimes referred to as mesh. A three-phase motor can run continuously in either star or delta. In star, the motor will run slower than in delta but it will not draw as much current or produce as much energy.

As an example:

A star connected motor has windings with a resistance of 10Ω per winding on a supply voltage of 400V.

The calculation to find the current in each phase is:

$$\frac{400}{\left(R \times \sqrt{3}\right)} = \text{current in each phase}$$

$$\frac{400}{\left(10 \times 1.732\right)} = 23\text{A}$$

This can be simplified for star as $\frac{230}{10} = 23\text{A}$.

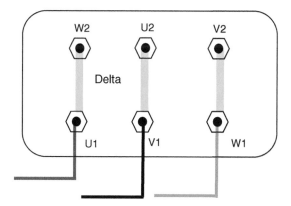

Figure 17.2

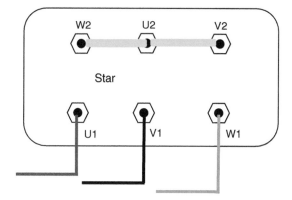

Figure 17.3

The same motor is connected in delta. The calculation to find the current in each phase is $\frac{400}{10} = 40$A.

As shown, the current in a motor connected in delta is 1.732 times greater than the current in the star connected motor.

Total power for both motors can be calculated as follows:

Power in watts = line voltage × current × $\sqrt{3}$
Star   400 × 23 × 1.732 = 15,934W
Delta   400 × 40 × 1.732 = 27,712W

As you can see, the power output for a motor connected in delta is considerably higher than that of a motor connected in star, but then so is the current use.

A star connected motor will have three ends of the windings joined together, and a delta connected motor will have six separate terminals. Let's say V2, U2 and W2 are joined (see Figure 17.3) and V1, U1 and W1 are connected to separate terminals.

If by any misfortune the ends of the windings are not identified because the markings have come off, it's no problem. Use a low resistance ohm-metre (there is no real need to zero the leads for this test but it becomes a habit after a while if you do a lot of testing). All you need to do is just pick a winding end and test that end to all of the others in turn until you get a resistance value. Once you have found a pair, mark them and then carry out the same procedure with the four remaining ends and, of course, you will end up with three pairs.

Providing the resistances of each winding is the same or very close, we can move on to the next step.

Using an insulation resistance tester set at 500V, test between each pair of windings, and then each winding to the case of the motor. All of the measured values should be very high, certainly above 1MΩ.

Values below this may be acceptable on motors that have not been used for a long period; the value may be low due to it being stored in a damp atmosphere. The winding resistances should all be the same but the insulation resistance may be low. In these cases it is a good idea to either store the motor in a warm and dry environment, or if you are in more of a hurry connect the motor and let it run for a few hours. This will allow the motor to warm up and dry out.

After a few hours, carry out the insulation test again. You will usually find that the resistance value will have increased; this will prove that it is just dampness.

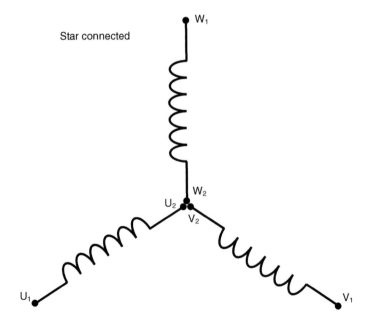

Polarisation testing of electric motors

Without going into the detail of the calculation, this is a test that is carried out to check for any deterioration of the windings due to perhaps a build-up of dust, dirt or moisture.

For this test you will need a good quality insulation resistance tester set at 500V DC or above, it is important that the windings are joined and shorted to the body of the motor before the test is carried out. If the motor is in use or has had other tests carried out on it, the windings should remain shorted to earth for 20 minutes or so to ensure that the PI test is a valid one.

To carry out the test, make sure that the motor windings and the body of the motor are not touching. Test at ≥ 500V DC for one minute and record the reading. Let's just say we we you have a measured value of 7MΩ.

Now repeat the test for a period of ten minutes and record the reading. Let's say 30MΩ.

Now we divide the ten-minute test by the one minute test:

$$\frac{30}{7} = 4.28$$

If the first insulation test provides a reading of 4GΩ or above, then it is not worth carrying out the PI test as the reading indicates that the motor is in good condition.

Table 17.1

Polarisation index results	
PI result	Condition of motor
<1.0	Hazardous
1.0–2.0	Bad condition
2.0–3.0	Usable
3.0–4.0	Good
4.0–5.0	Excellent

Chapter 18

Use of RCDs

Due to changes in the wiring regulations over the past 20 years or so the use of RCDs has become common practice. Before the 16th edition wiring regulations were introduced, the use of RCDs was pretty much limited to TT systems or circuits that could not meet the maximum Zs values.

There is no doubt that RCDs do save lives. However, occasionally than can be difficult to deal with because of things like nuisance tripping, high Zs values when testing and even the fear of damage when testing.

A basic RCD is a device that is used to detect earth faults. The RCD monitors the amount of current flowing in the line conductor and the neutral. Clearly, the amount of current flowing in each live conductor on a single circuit or a complete single-phase installation will be the same.

When an earth fault occurs and some of the current leaks to earth the result will be more current flowing in one live conductor than the other. Once the fault current reaches the trip rating of the RCD the device will automatically switch off. This is the operating principle of all RCDs. However, RCDs come in various forms.

All RCDs used in the UK should be manufactured to European standards and can be identified by looking at their BS EN numbers.

Some of the older RCDs will have only BS not BS EN numbers. BS 4293 was the British Standard for RCDs and this was superseded in 2012 by BS EN 61008.

It is a good idea to look at what the numbers stand for.

BS 4293 is an RCD that will detect an earth fault and switch off, but it will not switch off if there is an overcurrent. In other words, if the circuit is overloaded or there is a short circuit between live conductors it will only switch off for an earth fault.

This type of device is commonly found on slightly older properties, used as a main switch on the consumer unit or sometimes as a main switch before the consumer unit.

BS EN 61008-1 type RCDs are now listed as an RCCB. This indicates that it is a residual current circuit breaker without overload (Figure 18.1).

These are now used instead of the older BS 4293 RCDs, usually as a main switch (Figure 18.2).

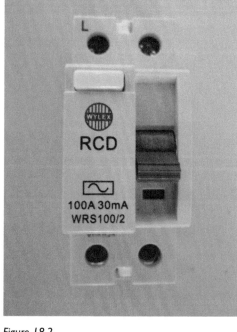

Figure 18.1   Figure 18.2

Zs values BS EN 61009-1 type RCDs are now listed as RCCBO. This indicates that it is a residual current circuit breaker with overload; these can be used to take the place of circuit breakers where RCD protection is required for individual circuits. These are common in domestic installations as the rating of them is quite limited, although they are also used in industrial and commercial installations (Figure 18.3).

In industrial and commercial installations, all classifications would need to be considered, although when using higher rated devices they may be referred to as a CBRs (circuit breakers with residual current protection).

Of course, RCD protection can be incorporated into individual socket outlets and these would be referred to as SRCDs. Where RCD protection is incorporated into fused connection units they would be referred to as SRCBO, as, of course, the fuse in the connection unit would provide overload protection.

There is often confusion when referring to RCDs and for that reason it is important to recognise the markings on them. In general they indicate the electrical characteristics of the device (Figure 18.4).

The first marking is I or A and this indicates the maximum load rating of the device.

Next, there will be IΔn. This shows the trip rating of the device.

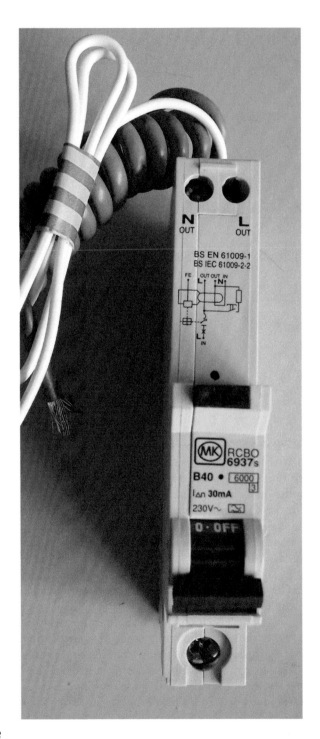

Figure 18.3

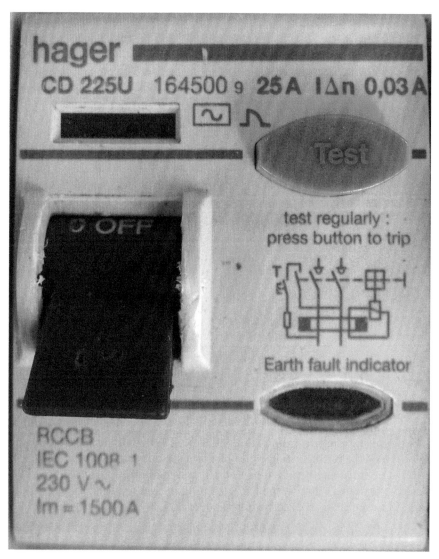

Figure 18.4

Some devices have an inbuilt time delay. This is because where two RCDs are connected in series it is not possible to discriminate as to which device will operate first, even if they have different trip ratings (I∆n). The only way that discrimination is possible between RCDs is to have an inbuilt time delay. This will prevent the device recognising a fault for a very short period of time. This will allow the chosen device to operate before the time

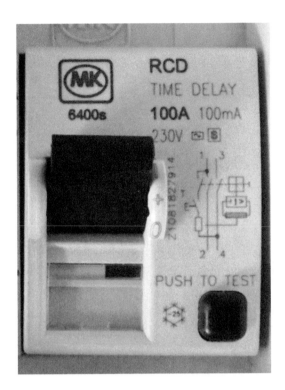

Figure 18.5

delayed device. Time delayed or selective type RCDs are marked with an S (Figure 18.5).

Recognising the trip rating (IΔn) of an RCD is vitally important, as the selection of the wrong rating could result in serious safety issues.

Devices are available in a range of trip and load ratings. It is pretty obvious that any RCD used should be able to carry the required load, but it is the trip rating that needs to be fully understood.

IΔn 2000mA

A high rated RCD could be up to 2000mA (2 seconds) and these would be used generally in temporary entertainment supplies or some industrial applications. These are often adjustable but adjustment must not be possible without the use of a tool (see regulation 531.2.10).

IΔn 300mA

This rating of RCD would normally be the upper limit used for fire protection (see regulations 422.3.9 and 532.1).

I∆n 100mA

Often used where the disconnection time required for a circuit cannot be met because the earth fault loop impedance (measured Zs) is not low enough to meet the requirements of BS 7671.

I∆n 30mA

Used where disconnection times cannot be met due to high Zs values but more commonly now used for *additional protection* for socket outlets (see regulation 411.3.3), circuits buried in walls with no mechanical earthed protection (see regulation 522.6) and circuits in bathrooms (see regulation 701.411.3.3).

I∆n below 30mA

These are more sensitive devices that can used for protection in areas where there is perhaps a greater risk of electric shock, such as work benches. They are possibly used with socket outlets in schools that may be used by students, and also in certain areas of hospitals where the level of fatal electric shock may be reduced due to health issues.

RCD types A, B and AC

As well as having different classifications such as RCCB and RCCBO, they are also rated as A, B and AC.

Type A will trip an AC and pulsating DC.

Type B will trip on AC and also pulsating and steady DC. This type of device should be used where RCD protection is required on photovoltaic systems.

AC will trip on AC and is suitable for most installations.

All of these devices will operate on sudden or slowly rising currents.

Testing of RCDs

The disconnection time of an RCD can only be tested using a dedicated RCD test meter.

Table 18.1a

RCDs to BS 4293 and RCD socket outlets to BS 7288	
Test at 50% of trip rating	Must not trip within 2 seconds
Test at 100% of trip rating	Must trip within 200ms
After the tests have been carried out, it is important that the functional test button is pushed and that the device trips. Check that there is a test label present, which indicates to the user the necessity for the test button to be pushed quarterly.	

Table 18.1b

RCCBs and RCBOs to BS EN 61008 and 61009

Test at 50% of trip rating	Must not trip within 2 seconds
Test at 100% of trip rating	Must trip within 300ms

After the tests have been carried out, it is important that the functional test button is pushed and that the device trips. Check that there is a test label present, which indicates to the user the necessity for the test button to be pushed quarterly.

Devices used for additional protection IΔn ≤ 30mA	
Test at five times the trip rating	Must trip within 40ms

Insulation resistance tests

There is no problem with carrying out a 500V insulation test through an RCD as they are manufactured to withstand this level of voltage. RCDs can, however, affect the value of the insulation results. If an insulation test through an RCD gives results lower than expected it may be necessary to disconnect the RCD and test the circuit individually.

Earth fault loop impedance tests with RCDs

When carrying out a high current earth fault loop test the test instruments inject a test current of up to 25A into the circuit. This, of course, will trip the RCD.

The older types of RCDs can be tested using a tester with what is called a 'D' lock mechanism. This type of test instrument uses DC to desensitise the trip coil, which will prevent it from operating during the earth fault loop test.

This type of instrument will not work on BS EN 61008 or 61009 RCDs, as the desensitising current will cause them to trip.

The easiest method used to carry out an earth fault loop test on RCDs is to use a low current tester that uses a test current of 15mA – this will not trip a healthy RCD.

There is an added problem with carrying out earth fault loop tests that have RCDs in the circuit, as the RCD, particularly RCBOs, often result in false high readings. There are, of course, methods that can be used to overcome the problem.

Method 1

- Take a Zs reading at the supply side of the RCD and make a note of it. Say, 0.39Ω.
- Take a Zs reading at the load side of the RCD and make a note of it. Say, 0.76Ω.

- Take a Zs reading at the furthest point on the circuit and make a note of it. Say, 0.95Ω.
- Now subtract the first reading from the second reading. The result will be the added resistance of the RCD.

0.76Ω – 0.39Ω = 0.37Ω

- Now subtract the resistance of the RCD from the Zs value measured at the end of the circuit and this will be the circuit Zs.

0.95Ω – 0.37Ω = 0.58Ω
Zs = 0.58Ω

Method 2

- Take a Zs reading at the supply side of the RCD.
- Measure $R_1 + R_2$ for the circuit and add the two resistances together. This will provide you with the calculated resistance.

Method 3

This is my preferred method but it does not suit everyone, although it is a very simple process and it can be carried out as follows:

- Isolate the circuit to be tested.
- Link phase and earth at the furthest point of the circuit using a lead with a crocodile clip on each end (Figure 18.6). If it is a socket outlet, then a plug top with earth and phase linked can be used. (*It is advisable to clearly mark the plug top.*)

Figure 18.6

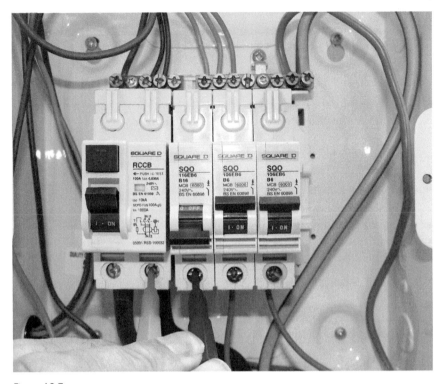

Figure 18.7

- Use a high current earth fault loop impedance test instrument.
- Place one probe (*black*) onto the isolated terminal of the circuit protective device.
- Place the other probe (*red*) onto the incoming phase of the RCD or main switch (Figure 18.7).
- Operate the instrument and record the result.
- This will be Zs for the circuit and the RCD will not have tripped.
- If your test instrument is a three-lead instrument, connect the black and green leads.
- The one problem with this test is if you don't remember to remove the line earth link, the RCD will keep tripping. But it will remind you that the link is still in place.

Chapter 19

Calculating the maximum Zs

There are three types of BS EN 60898 circuit breakers – Type B, C and D.

The main purpose of circuit breakers and fuses is to provide protection for the circuit cables. This means that they have to break current flow before any damage can occur to the cable insulation.

Damage can be caused to cable insulation when the cable heats up due to too much current passing through the conductor. There are two reasons why a cable could overheat. One is overcurrent, which occurs when the current passing through a conductor is greater than the rated current carrying capacity of the conductor. Overcurrent would occur when live conductors touch or a line conductor touches earth, resulting in a low resistance fault. This, of course, would result in a high current flowing in the circuit.

The other is known as overload current and is found in a circuit that has been designed correctly but is overloaded, possibly by too many current-using pieces of equipment being connected to a circuit or even a motor overloading or seizing.

Both of these conditions would have the effect of heating up a conductor, the difference being that overcurrent would heat the conductor up quickly and overload would heat the conductor gradually.

Circuit breakers and fuses have to be able deal with both of these conditions. A fuse is just a strip of metal or a piece of wire that will either melt instantly on overcurrent or slowly on overload. All protective devices must be able to withstand an overload current for a short period of time; on overload a piece of wire will take a while to get hot enough to melt.

Circuit breakers are designed to cope with both of these conditions.

On overcurrent that produces a sudden rush of current will operate a solenoid and switch off the device within 0.1 of a second.

Type B circuit breakers must operate within 0.1sec when a current of between three and five times the rated current of the device passes through it. A Type C circuit breaker must operate between five and ten times its rating and a Type D must operate between ten and 20 times its rating.

On overload the characteristics of circuit breakers must comply with the required British Standards, which are:

- Must not switch off within 1 hour when overloaded by 1.13 times its rating.
- Must switch off within 1 hour when overloaded by 1.45 times its rating.
- Must switch off within 2 minutes when overloaded by 2.55 times its rating.

To achieve this, a bi-metal strip within the device is used (Figure 19.1).

To calculate Zs for a circuit breaker is quite a simple task. Of course, it is always easier just to look up values on a chart but there are occasions when a chart is not available.

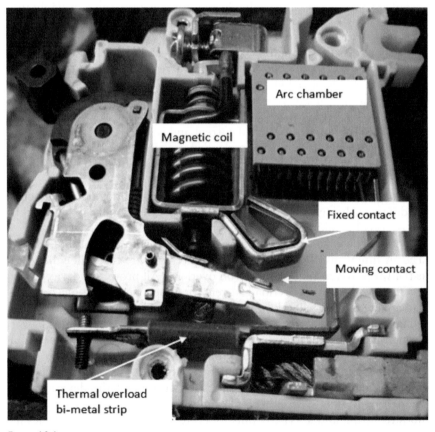

Figure 19.1

Step 1

Calculate the current required to operate the device.

As we have seen, a Type B will be five times its rating, Type C ten times and Type D 20 times.

Example: a 20A Type B circuit breaker must operate at a maximum of:

5 × its rating
5 × 20 = 100

This shows that the circuit breaker must operate instantly when a current of 100A is passed through it.

If the circuit breaker was a 20A Type C, then it would require:

10 × its rating
10 × 20 = 200

Step 2

Calculate the resistance that would allow the required current to flow. To carry out this calculation we require the value of the voltage that the circuit is operating at. For most installations this would be 230V. There is also a minimum voltage factor that must be applied; this is to take into account of voltage fluctuations which may occur in the supply system. The voltage factor is known as C_{min} and is 0.95. This C_{min} value has to be applied to the nominal supply voltage and the result divided by the current, which would cause automatic disconnection of the circuit breaker.

Example for a 20 A Type B:

$$\frac{230v \times 0.95C_{min}}{100} = 2.19\Omega$$

The maximum Zs value for a BS EN 60898 circuit breaker is 2.19Ω and this is the value that can be found in BS 7671:Amd 3.

This value is the one that should be used when carrying out a calculation for circuit conductor sizes, as it is the maximum value. However, when carrying out electrical testing this value becomes difficult to confirm without dismantling the circuit and measuring the $R_1 + R_2$ values.

We know that when conductors carry current they heat up. When they heat up they increase in temperature. When a copper conductor heats up it increases in resistance by 2% for every 5°C change in temperature. When we carry out an earth loop impedance test on a circuit we generally have no idea of the temperature of the conductors. This is because the conductors could be in containment with other circuits. The other circuits could be under load and hot. Some of this heat would be transferred to the circuit under test, which would, of course, increase the temperature of the conductors. The room temperature may also be high and this would also have an effect on the conductors under test.

If we were to carry out an earth fault loop impedance test on a circuit protected by a 20A Type B circuit breaker and the measured value was 2Ω, we could assume that the circuit was perfectly acceptable. Unfortunately, we may be wrong because we do not know the temperature of the conductors during the test.

To compensate for these temperatures we can use a method/calculation known as the rule of thumb. All that is required is for us to multiply the maximum Zs by a figure of 0.8 and compare the result to the measured value.

The measured value of Zs must be lower than the corrected value of Zs. Example: $2.19 \times 0.8 = 1.75Ω$.

Now if we compare the corrected value of 1.75Ω to the measured value of 2Ω we can see that the corrected value is less than the measured value. This indicates that the circuit is unsatisfactory because the resistance will rise to a value of above 2.3Ω if the conductor were ever to reach its maximum operating temperature.

Fuses

As described earlier, all circuit breakers when installed correctly will operate on overcurrent within 0.1sec, although the requirements of BS 7671 are that circuits on TN systems with a current rating of up to and including 32A need to operate within 0.4sec, and circuits above 32A must operate within 5sec. For TT systems the disconnection times are 0.2sec for circuits up to and including 32A.

Because the overcurrent part of a circuit breaker is operated by a solenoid that is activated by a high current, the 0.1sec is simple to predict. Where fuses are involved the overcurrent has to melt the fuse element. This will take slightly longer and on occasion can take the full 0.4sec. Of course, this depends on the magnitude of the fault current.

There are, of course, exceptions. Distribution circuits (sub mains) can be permitted to have an extended disconnection time of 5sec for TN systems and 1sec for TT systems. Clearly, this will only apply to fuses, as all circuit breakers will operate within 0.1sec providing the fault current reaches the required values. Lower fault currents will result in the device operating time being extended to probably 10sec or more.

RCBOs

The difference between an RCBO and a circuit breaker is that an RCBO contains a residual current detection device; this detects an imbalance between the live conductors when a fault to earth occurs. The overload and short-circuit parts of an RCBO operate in exactly the same way as a circuit breaker, with the disconnection times being the same. It is the earth fault current part of the device that is different. The product standard for RCBOs and RCDs to BS EN 61008-1 and BS EN 61009-1 state that they must operate within a maximum of 300ms (0.3sec) at their rated trip current.

Compliance with BS 7671 requires that circuits up to and including 32A connected to a TT system must disconnect under fault conditions within 200ms (0.2sec). If the fault is between live conductors, this will not present a problem as the solenoid part of the device will operate within 0.1sec. It is when the fault is an earth fault that problems may arise when an RCBO is installed in a TT system.

As previously explained, the fault current part of a circuit breaker requires a high inrush of current to operate it quickly. This is fine where the fault is between live conductors on TN and TT systems, as usually the short current is quite high due to the low resistance of the conductors.

Problems arise when we are required to connect to TT systems due to the fact that the Zs values are normally quite high as we have to rely on earth electrodes. Most earth electrodes will provide quite high Ze values; these are usually overcome by the use of residual current devices. RCBOs are circuit breakers thar contain a residual current device. This, of course, means that they can be used to provide protection where Zs values exceed the permitted maximum Zs.

As we have seen, the requirements for TT systems are that a protective device must operate within 0.2sec under fault conditions and that the product standard for RCDs states that they must operate within 0.3sec at the trip current rating of the RCD. This often causes confusion, particularly where RCBOs are used where the circuit has high Zs values.

The product standard for RCDs and RCBOs states that, although they have to disconnect within 300ms at their rated trip current value, they also have to trip within 150ms at twice their current rating. When using these devices for earth fault protection on TT systems compliance with the disconnection times permitted in BS 7671 is achieved because it is accepted that any fault currents will be significantly higher than the residual operating current of the RCD or RCBO.

Chapter 20

Fire protection

With the increased use of RCDs in electrical installations, consideration must be given to the type of circuit being installed. We are aware that buried cables must meet certain requirements, which often requires the use of RCDs.

Clearly, the use of RCDs alongside fire alarm systems is not a good idea.

Approved document B along with other British Standards such as BS 5839-1 and BS 5839-6 forbid the use of RCDs in fire alarm supplies, unless it cannot be avoided to comply with the requirements of BS 7671.

As we know, these British Standards do not stand alone and have to be complied with alongside all other British Standards; just as with the wiring regulations the installer cannot be selective, all parts of the standards need to be met. A list of the British Standards that may affect installations can be found in Appendix 1 of BS 7671 Requirements for Electrical Installations.

A TT installation would be a good example of this. Where the use of RCDs cannot be avoided, the device must be completely segregated from the rest of the installation to ensure that a fault in the general installation will not result in a loss of supply to the fire detection system.

Although nuisance tripping is pretty much a thing of the past these days, an RCD in the system could be seen as an unnecessary risk, which needs to be avoided.

The use of time delayed RCDs would not be a suitable solution where used as a main switch on a board containing RCBOs, as many RCBOs are only single pole. Although the device will trip on a N–E fault the fault will still remain and trip the time delayed device.

A better option would be just to use RCBOs for all of the circuits and a simple on/off main switch, or even the use of double insulation may be considered.

Providing the circuit is under supervision and is only used to supply an all-insulated alarm panel, cables installed in an all-insulated containment system that is surface mounted would be a suitable solution.

It should also be remembered that, because the fire alarm system cannot overload, overcurrent protection is not required and cables with a lower rating than the protective device can be installed if required. Short circuit and earth fault protection must, of course, remain in place.

In general, the use of RCDs alongside fire alarms or any fire detection system should be avoided wherever possible. As already mentioned, where TT systems are involved the two options are to use RCDs or preferably an all-insulated system.

Sometimes, however, the client will require the supply cable to the alarm supply to be buried within the building structure. As we know, all buried cables need some kind of additional protection. The use of RCDs would often be the simplest method for protecting most buried cables.

Where the cable is for fire protection and the use of RCDs needs to be avoided, consideration should be given to wiring the alarm supply circuit in FP200 gold or flexi shield cable, as this is by far simpler to use for this sort of job than, say, steel wire armoured cable or mineral insulated cable.

Fire protection in dwellings

Approved document B: Fire safety, Volume 1 sets out the requirements for fire protection in dwellings.

A dwelling is a residential unit that is occupied by a single person, people who are living together as a family, or a residential unit occupied by not more than six people living together in a single household. This also includes a residential building where care is provided for up to six people.

Flats or buildings containing flats are not classed as dwelling houses.

As a minimum, dwelling houses with a floor area of up to 200m² per storey should be provided with a fire detection system to at least Grade D. This means that each alarm (smoke or heat) must be mains powered with an integral backup in case of mains failure. Where there is more than one alarm, they must be interconnected and connected to the same circuit.

In a typical installation in a house where the floor area does not exceed 200m² on any storey and the kitchen is separated from the rest of the house by a door/s, there must be one smoke alarm on each hall or landing and one in the master bedroom.

Where the kitchen is not separated from the rest of the house there must be a heat or smoke alarm in the kitchen and a smoke alarm in each hall or landing.

For both scenarios the hall or landings would be classed as circulation spaces. The smoke alarms should be sited within 7.5m of every room and a minimum of 300mm from any luminaire. In some cases, more than one alarm may be required in a single circulation space.

The electrical supply can be a dedicated supply, in which case it is a good idea to clearly label the circuit. However, a better source for the mains supply would be another circuit, such as a lighting circuit, which is regularly used. This will, of course, provide a good indicator of any power failure.

Larger houses with a floor area exceeding 200m² have to be fitted with a fire detection and alarm system to a minimum of Grade B.

Bungalows with a floor area exceeding 200m² can have a Grade D installation.

A Grade B system is a fire detection and fire alarm system that incorporates fire detectors and alarm sounders. Smoke and heat alarms would not fall into this category.

The mains supply must be a dedicated circuit that is used only for the fire detection system. It must also have a standby system that will maintain the system for a minimum of 72 hours. After 72 hours there must be enough energy left to supply the alarm load for a minimum of 15 minutes.

The mains supply must be clearly labelled 'FIRE ALARM: DO NOT SWITCH OFF' and the wiring system for the alarm must be fire resisting. This can be achieved by using fire-resisting cables or a fire-resisting containment or support system.

When using a Grade A system it is a requirement that all cables supplying a fire alarm system are segregated from other parts of the electrical installation. This can be achieved by using segregated trunking (Figure 20.1) with sheathed multi-core cables, or by installing sheathed multi-core cables that are physically separated from other parts of the installation.

Apart from grades of systems, there are also categories of system. Generally, these are made up of two letters and a number.

The letters indicate the following:

- L is protection of life.
- P is for protection of property.
- D is for dwelling.

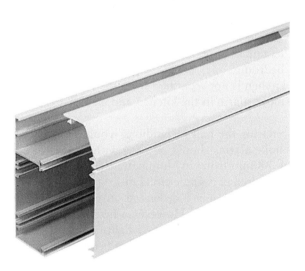

Figure 20.1

Category	Description
LD	The fire detection system intended for the protection of life in dwellings. LD is used to indicate dwellings, whereas L is for protection of life in any building
LD1	This system has detectors in all circulation spaces that form escape routes, and in all rooms where a fire may start. This excludes bath/shower rooms and WCs
LD2	This system has detectors in all circulation spaces that form escape routes, and in all rooms/areas that have a high fire risk. These rooms would be kitchens, living rooms, etc.
LD3	This system has detectors in all circulation spaces that form escape routes
PD	This would be for protection of property. PD indicates dwellings, and where the letter P is used it is intended for protection of all types of property
PD1	Detectors fitted in all rooms where a fire may start, excluding bath/shower rooms and WCs
PD2	Detectors fitted in rooms that are deemed to have a higher risk of fire

Escape routes

Fire escapes must be identified. An escape route can be any route from any building or room, and care must be taken when installing containment systems and cables to ensure that in the event of a fire they do not collapse and block the route.

Consideration must be given to the type of system; a metallic system would simply require the fixings to be suitable and not be affected by the high temperatures that will be present in a fire.

Where plastic containment systems are used or cables are fixed directly to the surface, it is important that the cables are secured to prevent them from collapsing.

The method used to secure cables installed in trunking is to fix a purpose-made clip (Figure 20.2) inside the trunking, which is secured through the

Figure 20.2

Figure 20.3

trunking using a screw. Wherever possible, it is preferable to use a masonry screw that is fixed directly into the brick or concrete (Figure 20.3). Due to the very high temperatures that would be present in a fire, it is possible that a plastic plug holding the screw may soften and eventually allow the screw to pull out.

The major problem with cables and containment systems collapsing is that the situation would put any firefighters at risk, not particularly when they enter the building but when they try to get out of the building or room. There have been fatalities and serious injuries caused by the cables getting tangled in the firefighters' breathing apparatus.

These requirements have a major influence on the way that we use plastic containment systems. Many installations, particularly rewires carried out in flats, require that the circuits are installed in surface trunking.

As any exit from a room may be classed as an escape route, any parts of the installation that may collapse in a fire must be secured. This, of course, means that the cables must be secured correctly. These rules also apply to data cables, phone cables and alarm systems.

Index